T0214262

Lecture Notes in Computer Science 11720

Massimiliano Albanese · Ross Horne ·
Christian W. Probst (Eds.)

Graphical Models for Security

6th International Workshop, GraMSec 2019
Hoboken, NJ, USA, June 24, 2019
Revised Papers

 Springer

Editors
Massimiliano Albanese (ID)
George Mason University
Fairfax, VA, USA

Ross Horne (ID)
University of Luxembourg
Esch-sur-Alzette, Luxembourg

Christian W. Probst
Unitec Institute of Technology
Auckland, New Zealand

ISSN 0302-9743 ISSN 1611-3349 (electronic)
Lecture Notes in Computer Science
ISBN 978-3-030-36536-3 ISBN 978-3-030-36537-0 (eBook)
https://doi.org/10.1007/978-3-030-36537-0

LNCS Sublibrary: SL4 – Security and Cryptology

This Springer imprint is published by the registered company Springer Nature Switzerland AG
The registered company address is: Gewerbestrasse 11, 6330 Cham, Switzerland

Preface

This volume includes all the contributions made to the 6th edition of the International Workshop on Graphical Models for Security (GraMSec 2019). As in the previous four editions, GraMSec 2019 was held as a workshop co-located with the IEEE Computer Security Foundations Symposium (CSF), which was in its 32nd edition (CSF 2019). Both events were hosted at the Stevens Institute of Technology campus located at 1 Castle Point Terrace in Hoboken, New Jersey, USA. GraMSec 2019 was held on June 24, 2019, as a pre-conference workshop.

This edition of GraMSec reflected on the diversity of graphical approaches for analyzing the security of systems. The traditional priority area of GraMSec has been the area of formal methods underpinning graphical models such as attack trees and attack graphs. Attack trees can be used for structured reasoning about the sub-goals of attackers, while attack graphs typically capture the dynamics of an attack surface. As models mature and new applications are explored, GraMSec is taking an increasingly broader view of what constitutes a graphical model in the analysis of systems security. Indeed, taking such a broader view can be regarded as a necessary step in the effort to continually adjust graphical methods to meet the needs of security practitioners and facilitate the transition to practice of research outcomes.

We had the honor to open the workshop with an invited talk by George Cybenko, the Dorothy and Walter Gramm Professor of Engineering at the Dartmouth's Thayer School of Engineering. Prof. Cybenko provided insights into a research direction where control flow graphs, a classical graphical method for program analysis, could be inferred from radio frequency emissions of microprocessors.

Other graphical methods explored in this workshop, beyond attack trees and graphs, included: bow-tie diagrams for modeling causal dependencies surrounding a security risk, and project management enhancements for the EBIOS cyber security methodology. Some papers also considered a graphical subject matter such as social network graphs. However, the mainstay of the research presented remains in the field of attack trees, as is traditional for GraMSec, where papers both extended existing attack tree methodologies and developed compelling case studies.

This year, we received 15 submissions, of which 8 were accepted as regular papers. These proceedings were published after the event, giving authors the opportunity to revise and enhance their manuscripts based on feedback received at the workshop. These proceedings also feature two invited papers from our keynote George Cybenko and from Olga Gadyatskaya and Sjouke Mauw. Papers received an average of 3.5 reviews in a single-blind review process, with each Program Committee member reviewing two papers on average.

We thank the Program Committee members for their timely responses to review requests, and the authors for their sterling contributions. We also extend a special thanks to Barbara Fila and Fabio Persia, who volunteered to chair workshop sessions, thus contributing to the smooth running of the workshop. Finally, we thank Dominic

Duggan, the general chair of CSF 2019, who took the responsibility of coordinating all the logistics for the main conference and co-located workshops. We look forward to future editions of this workshop series.

October 2019

Massimiliano Albanese
Ross Horne
Christian W. Probst

Organization

General Chair

Christian W. Probst   Unitec Institute of Technology, New Zealand

Program Committee Chairs

Massimiliano Albanese   George Mason University, USA
Ross Horne   University of Luxembourg, Luxembourg

Steering Committee

Sushil Jajodia   George Mason University, USA
Barbara Kordy   INSA Rennes, Irisa, France
Sjouke Mauw   University of Luxembourg, Luxembourg
Christian W. Probst   Unitec, New Zealand
Ketil Stølen   SINTEF Digital and University of Oslo, Norway

Program Committee

Ludovic Apvrille   Télécom ParisTech, France
Zaruhi Aslanyan   Alexandra Institute, Denmark
Stefano Bistarelli   Università di Perugia, Italy
Hasan Cam   U.S. Army Research Laboratory, USA
Nora Cuppens-Boulahia   IMT Atlantique, France
Harley Eades III   Augusta University, USA
Olga Gadyatskaya   SnT, University of Luxembourg, Luxembourg
Guy McCusker   University of Bath, UK
René Rydhof Hansen   Aalborg University, Denmark
Jin B. Hong   The University of Western Australia, Australia
DongSeong Kim   The University of Queensland, Australia
Barbara Kordy   INSA Rennes, Irisa, France
Sjouke Mauw   University of Luxembourg, Luxembourg
Per Håkon Meland   SINTEF ICT, Norway
Guozhu Meng   Chinese Academy of Sciences in Beijing, China
Vivek Nigam   fortiss GmbH, Germany
Andreas Lothe Opdahl   University of Bergen, Norway
Noseong Park   George Mason University, USA
Stéphane Paul   Thales Research and Technology, France
Sophie Pinchinat   INSA Rennes, France
Saša Radomirovic   University of Dundee, UK

Rolando Trujillo Rasúa	Deakin University, Australia
Paul Rowe	The MITRE Corporation, USA
Giedre Sabaliauskaite	Singapore University of Technology and Design, Singapore
Ketil Stølen	SINTEF, Norway
Sridhar Venkatesan	Vencore Labs, USA

Publicity Chair

| Ibifubara Iganibo | George Mason University, USA |

Additional Reviewers

Tim Willemse
Michel Hurfin
Aleksandr Lenin
Francesco Santini
Valérie Viet Triem Tong

Contents

Invited Papers

Graph Models in Tracking Behaviors for Cyber-Security

George Cybenko$^{(\boxtimes)}$

Thayer School of Engineering, Dartmouth, Hanover, NH 03750, USA
gvc@dartmouth.edu

Abstract. This chapter briefly summarizes recent research on the problem of inferring security properties of a computation from measurements of unintended electromagnetic emissions from the processing system on which the computation is being executed. The particular approach described involves two ingredients: (i) signal processing and machine learning to map observed analog measurements to program segments; and (ii) the program's control flow structure which constrains the legitimate transitions between program segments. In particular, the control flow logic of a program can be represented as a *control flow graph* (CFG) that summarizes possible execution paths and control flows in terms of transitions between basic blocks of the executable. In other words, the ultimate goal of this work is to track the behavior of an execution using unintended electromagnetic emissions. We describe various control flow graphs properties that impact the extent to which valid execution of a program can be monitored and subsequently used for program classification and anomaly detection. Suggestions for future work on graph models are described as well.

Keywords: Control flow graphs · Electromagnetic emissions · Cyber security

1 Basic Problem Statement

Many operational computing systems use processors with minimal memory and/or processing power yet are susceptible to cyber attacks nonetheless. Because of limited resources, it may not be possible to add intrusion detection logic to the computational platform. Such situations can arise in both modern Internet of Things (IoT) and legacy applications. Moreover, in legacy and certain safety critical applications, it may not even be possible to modify the code or processing system due to regulatory or sustainability factors.

Short of being able to modify the code or the processing platform, it may be possible to monitor the execution of the system using unintended electromagnetic emissions, specifically radio frequency signals that are generated by the electrical activity in the processor executing instructions of the program of interest.

Detailed descriptions of the underlying physics of the radio signal generation and antenna technology used, as well as the associated signal processing and

© Springer Nature Switzerland AG 2019
M. Albanese et al. (Eds.): GraMSec 2019, LNCS 11720, pp. 3–6, 2019.
https://doi.org/10.1007/978-3-030-36537-0_1

machine learning of the received signals, can be found in several publications describing those aspects of the effort [3,5,6].

A key point in the approach is that received signals are processed and ultimately classified into a finite discrete set of classes that we call "colors." These colors correspond to program segments that are being executed so we observe a time series of colors or classes, and seek to use that time series to make assertions about the underlying program execution.

There are many challenges in these types of problems. Among them are:

- Which subset of radio frequencies are best suited for identifying execution properties of a program running on a specific type of processor? (Millions of frequencies are possible to collect but only dozens can realistically be used for real time inferencing)
- How granular can the program segments be made and how granular do they have to be made for a particular use case? (Instructions, basic blocks, subroutines/functions, etc.)
- What specific program execution properties need to be detected for a specific use case? (Control flow integrity, instruction execution integrity, overall program characterization, etc.)
- What is an acceptable latency of detection for a given use case? (That is, the time when a specific condition occurs to the time when that condition is detected reliably.)

2   Graph Theoretic Aspects

The graph theoretic model of the problem can be reduced to various variants of the so-called "Graph Observability" problem as introduced by Jungers and Blondel in 2011 [8]. Consider processed and classified observations of RF signals as being assigned colors. The control flow graph of a computation has nodes that are colored by the RF emissions those node produce. As the computation proceeds, a time series of colors is observed.

The observability property of a colored graph has to do with whether sequences of observed node colors will eventually identify the last node uniquely. Because there is rarely a one-to-one mapping between nodes (program segments) and unique colors, this can be a complex property to check for large graphs.

Prior work has demonstrated that observability is related to other types of inference problems on colored graphs, resulting in a hierarchy of graph types with differing tracking properties [1,5].

Observable graphs are the most desirable because after observing some finite number of colors, it is possible to infer which node of the graph the last observation was produced by, irrespective of which node or color the observations began with.

Unifilar graphs [7,10] are colored graph analogs to Deterministic Finite Automata in that once we know which node we are at, we will deterministically be able to uniquely infer all future states. All observable graphs are unifilar.

The next weaker class of graph models studied are the so-called "trackable graphs" which have the property that given a color observation sequence of length T, the number of possible trajectories (hypotheses) that could have generated that color sequence is polynomial in T as $T \to \infty$ [4].

None of these classes of graph have stochastic aspects as described above. Adding noise and uncertainty to the RF signal to color observation and allowing probabilistic transitions between nodes (that is, basic blocks or function calls within a program) creates a Hidden Markov Model [9]. Such models do not seem appropriate however because nodes within a control flow graph are not actual *states* of an executing program. Moreover, transitions between basic blocks are rarely Markovian.

The most recent results about colored graphs and their properties can be found in several recent publications [1, 2, 5].

Several interesting challenges remain in this general area, including:

– Dealing with very large computations (it has been empirically established that average basic blocks are 10 instructions or fewer in size so that programs with millions of instructions will have hundreds of thousands or millions of basic blocks and therefore nodes in their control flow graphs);
– Designing computations (that is, their control flow graphs and associated coloring) so that they can be highly observable to make such analyses more effective;
– Dealing with multi-threaded, multi process computations and systems in which context switching is permitted.

Acknowledgements. This work summarized here was the result of many fruitful and enjoyable collaborations with my colleagues and co-authors including: Mark Chilenski, Valentino Crespi, Isacc Dekine, Guofei Jiang, Piyush Kumar, Gil Raz and Yong Sheng.

References

1. Chilenski, M., Cybenko, G., Dekine, I., Kumar, P., Raz, G.: Observability properties of colored graphs. arXiv preprint arXiv:1811.04803 (2018)
2. Chilenski, M., Cybenko, G., Dekine, I., Kumar, P., Raz, G.: Observability properties of colored graphs. Submitted for publication (2019)
3. Cleveland, C., Chilenski, M., Dekine, I., Kumar, P., Raz, G.: Microsystem identification and fingerprinting using RF side channels. Technical report, Systems and Technology Research Woburn United States (2019)
4. Crespi, V., Cybenko, G., Jiang, G.: The theory of trackability with applications to sensor networks. ACM Trans. Sens. Netw. (TOSN) **4**(3), 16 (2008)
5. Cybenko, G., Raz, G.M.: Large-scale analogue measurements and analysis for cyber-security, vol. 3, p. 227. World Scientific (2018)
6. Dekine, I., Chilenski, M., Cleveland, C., Kumar, P., Raz, G., Li, M.: Information theoretical optimal use of RF side channels for microsystem characterization. In: Cyber Sensing 2019, vol. 11011, p. 1101108. International Society for Optics and Photonics (2019)

7. Jiang, G.: Weak process models for robust process detection. In: Sensors, and Command, Control, Communications, and Intelligence (C3I) Technologies for Homeland Security and Homeland Defense III. Proceedings of SPIE, vol. 5403 (2004)
8. Jungers, R.M., Blondel, V.D.: Observable graphs. Discrete Appl. Math. **159**(10), 981–989 (2011)
9. Rabiner, L.R.: A tutorial on hidden Markov models and selected applications in speech recognition. Proc. IEEE **77**(2), 257–286 (1989)
10. Sheng, Y., Cybenko, G.V.: Distance measures for nonparametric weak process models. In: 2005 IEEE International Conference on Systems, Man and Cybernetics. IEEE (2005)

Attack-Tree Series: A Case for Dynamic Attack Tree Analysis

Olga Gadyatskaya and Sjouke Mauw[✉]

CSC/SnT, University of Luxembourg, Esch-sur-Alzette, Luxembourg
olga.gadyatskaya@uni.lu, sjouke.mauw@uni.lu

Abstract. Attack trees are a popular model for security scenario analysis. Yet, they are currently treated in the literature as a static model and are not suitable for dynamic security monitoring. In this paper we introduce *attack-tree series*, a time-indexed set of attack trees, as a model to capture and visualize the evolution of security scenarios. This model supports changes in the attack tree structure as well as changes in the data values. We introduce the notion of a *temperature function* as a special type of attribute that expresses the importance of change in the data values. We also introduce a *consistency* predicate on attack trees to allow inter-relating the evolving scenarios captured as attack trees. Finally, we discuss various application scenarios for attack-tree series and we demonstrate on a case study how the proposed ideas can be implemented to visualize historical trends.

1 Introduction

Today, organizations face unprecedented difficulties to stay secure. Cyber-weaponry becomes more and more commoditized, new vulnerabilities are reported every day, and attack surfaces become ever more complex and interdependent. To win this game against cyber-adversaries, security officers and analysts need to be able to access threat data over time, allowing them to pick up relevant trends and to proactively upgrade security [33].

Useful threat data can come in *quantitative* or *descriptive* formats. Frequency of attacks, cost of exploit kits, and number of lost data records are examples of the former category, while malicious IP addresses, indicators of a new adversary, recent zero-day attacks are examples of the latter one. Access to the latest threat data can be arranged, e.g., via subscriptions to threat feeds and monitoring of the relevant media channels [21,33]. Generally, organizations also collect a lot of threat data internally, from a security information and event management (SIEM) system.

All these threat data types and sources are highly dynamic, but not all threat modelling and analysis techniques applied in organizations are able to accommodate analysis of such data over time.

In particular, in this paper we focus on *attack trees*. Also known as threat trees, they are a popular modelling notation for expressing security threat scenarios. Attack trees are widely applied in organizations to capture anticipated

© Springer Nature Switzerland AG 2019
M. Albanese et al. (Eds.): GraMSec 2019, LNCS 11720, pp. 7–19, 2019.
https://doi.org/10.1007/978-3-030-36537-0_2

attack scenarios [12, 24, 30], to facilitate brainstorming [6], and to estimate different aspects of the considered threats via quantitative analysis [3, 16].

Yet, existing attack tree design methodologies and quantitative analysis (QA) techniques focus exclusively on *static scenarios*, in which each tree node and data value is fixed. There is no systematic way to detect trends in dynamic security scenarios on attack trees or to visualize the evolution of the security posture with respect to the considered threat model. This is an unfortunate omission. Indeed, an organization may have access to very relevant threat data over time, but it does not have a way to systematically link this valuable data to the attack tree model.

We argue that there is a case for enhancing attack trees for dynamic, evolving scenarios. In this position paper, we propose an approach to systematically capture the dynamic nature of both facets of the threat data, quantitative and descriptive, in an attack tree. We make the notion of *time* explicit in the definition of attack trees by developing an idea of *attack-tree series* that are not static but accommodate a sequence of attack trees.

We develop the first formalization of attack-tree series (ATSs) in this paper. In our formalization, we allow quantitative threat data values assigned to tree nodes to change over time. We also propose to capture the *importance* of the change as a *temperature* function that enables analysis and monitoring of historical trends. Our methodology supports scenarios in which not only the data, but also the attack tree itself evolves over time. This allows for a dynamic description of the various threats and their relations. In order to develop a consolidated view on an evolving series of attack trees, we introduce the notion of *consistency*. Consistency of a series of attack trees makes it possible to relate the data values attributed to the nodes of the various attack trees to each other.

To better highlight the applicability of the new attack-tree series, we also discuss potential use cases for them and demonstrate their usefulness for highlighting historic trends with an attack tree capturing automated teller machines fraud.

2   Related Work

In short, an attack tree represents a collection of attack scenarios. The *main goal* of the adversary common to all these attacks is represented by the *root* of this tree. The root is iteratively *refined* into more detailed attack components. The refinement process uses well-defined decomposition operators, typically OR and AND, and it continues until the analysts are satisfied that the nodes at the lowest level are atomic attack steps. These unrefined nodes are called *leaves* of the tree.

Attack trees have been introduced by Amoroso [2] and Salter, Saydjari, Schneier and Wallner [29], and formalized by Mauw and Oostdijk [22]. The basic attack tree model has been further extended into attack-defense trees [15] and attack-countermeasure trees [28] that both feature nodes representing security controls.

There exist several approaches to design an attack tree. An expert, or a group of experts, may design a tree manually, by analytically and iteratively considering all relevant attack developments. This is the traditional approach in the threat modelling literature [30, 31]. The manual work may be facilitated by relevant knowledge, e.g., from industry-specific catalogues of threats or threat ontologies [6]. Recently, automated and assisted attack tree generation techniques have emerged [10, 13, 14, 26, 34]. However, all these approaches work with static scenarios, and they do not take into account potential evolution in the considered threat structure.

The most popular quantitative analysis technique for attack trees is the *bottom-up computation approach*, in which leaf nodes are assigned values representing *attributes* (e.g., cost or probability of the corresponding atomic attack step). These values are then propagated bottom-up using the attribute rules for the corresponding decomposition operator [22, 30]. Various attributes that can be computed on an attack tree with the bottom-up computation approach are proposed by Bagnato et al. [3] and Kordy et al. [16]. For example, one can estimate probability of a particular threat, cost of an attack for the adversary or for the defender, satisfiability of attacks, time till successful attack, and many other parameters [3, 16]. These quantitative analysis techniques are used, for instance, to perform security risk analysis [18, 25] and to make decisions about security investments [8]. There exist other QA techniques for attack trees, usually involving the transformation of an attack tree into another model, e.g., timed automata [7, 18].

As we mentioned, existing QA approaches for attack trees work with static data values. Currently, if an organization wants to explore trends in the data values used in an attack tree, it needs to decouple visualization of security data values from the attack tree, for example, in a separate dashboard. Yet, this solution neglects the analyst's intuition behind the attack tree design, and could possibly hinder the detection of higher-level trends.

To the best of our knowledge, there are only few works studying attack tree visualizations. ADTool [9, 17] and SecurITree [1] are tools for manual design and quantitative analysis of attack trees, but they do not offer extensive visualization capabilities besides showing static attack trees. Li et al. [20] have developed an approach to visualize complex scenarios with attack trees. Their work focuses on very large attack trees, helping the analyst to quickly identify interesting areas (e.g., the most probable attack scenarios) by highlighting them visually. Yet, this technique has been developed for static scenarios. It is thus the goal of our work to introduce an approach to model dynamic security scenarios with attack trees and to perform quantitative analysis of such scenarios.

Outside the attack tree literature, security visualization is a well-developed research area, and dynamic data visualisation techniques for security risks and attacks have been studied in, e.g., [11, 19, 23, 27, 32]. These studies on cybersecurity visualisation, as well as Li et al. [20], agree on the need to support the analyst by directing their attention to the *important* areas in the model. In the next section we outline our proposal for modelling dynamic scenarios with attack trees enhanced with a novel type of attributes that capture the importance of data dynamics.

3 Attributes on Attack-Tree Series

In this section we formally define an attack-tree series as a sequence of time-indexed attack trees. We also define the valuation of an attribute on an attack-tree series and consider a specific type of attribute, which we call *temperature*. Because the proposed extension is independent of the chosen attack tree semantics, we will provide an intuitive interpretation only.

3.1 Attack-Tree Series

We consider attack trees constructed from leaf nodes and two types of internal nodes (AND and OR). Following, e.g., [5], we will assume that all nodes of the attack tree are labeled.

Definition 1 (Attack tree). *Let $\mathcal{L}$ be a set of labels. An attack tree is an expression generated by the following grammar (for $\ell \in \mathcal{L}$):*

$$t :: = \ell \mid \mathtt{OR}(t, \dots, t)_\ell \mid \mathtt{AND}(t, \dots, t)_\ell.$$

We say that an attack tree has *unique labels*, if all labels occur at most once in that attack tree. In this paper we will only consider attack trees that have unique labels. By *labels(t)* we denote the set of labels occurring in attack tree t.

An attack-tree series describes the evolution of an attack tree over time. In order to have a consistent view on the development of the individual nodes in the attack tree, we will define a consistency property.

Definition 2 (Consistent trees). *We say that two attack trees t and t' are* consistent *if the following three conditions are fulfilled:*

1. *The root nodes of t and t' have the same label.*
2. *If two non-root nodes of t and t' have the same label l_1, then their parent nodes must have the same label l_2.*
3. *If two non-leaf nodes of t and t' have the same label l_1, then they must have the same type (i.e. OR, resp. AND).*

A sequence (or set) of attack trees is consistent *if all its constituent attack trees are pairwise consistent.*

This notion of consistency guarantees that labels occurring in multiple incarnations of a developing attack tree always relate to the "same" node of the tree. Formulated differently, assume that we have two attack trees t and t' of which the root nodes have the same label, and assume that the intersection of the labels of these trees is L, then the subtree of t with labels from L is identical to the subtree of t' with labels from L.

Definition 3 (Attack-tree series). *Let Δ be a discrete time domain. An attack-tree series $(t_\delta)_{\delta \in \Delta}$ is a consistent sequence of attack trees indexed over Δ.*

An example of a consistent attack-tree series is shown in Fig. 1. Ignoring the attributes and temperature function (i.e. the numbers and colours), this figure shows an evolving attack tree at time points $\delta = 1, \ldots, 5$. Each attack tree in the series contains unique labels. It is easy to verify that every pair of attack trees in this series is consistent. For instance, the first two trees in the series are consistent because (1) their root nodes have the same label a, (2) corresponding non-root nodes in the two trees have parents with the same label (the parent of b in both trees is a), and (3) all non-leaf nodes with the same label have the same type (e.g. a is an OR node in both trees).

This figure also illustrates the rationale behind restricting Condition 3 in Definition 2 to non-leaf nodes. The reason for this restriction is that during the evolution of an attack tree an analyst may recognize substructure in an attack step that was first considered atomic. In the attack tree, this would mean that the node representing the atomic attack step evolves from a leaf node into an internal node of type OR or AND. As we want to relate the original leaf node to the new internal node, we only require corresponding types for non-leaf nodes, as expressed in Condition 3. This is illustrated by the first two attack trees in Fig. 1, where leaf node d changes into an internal node.

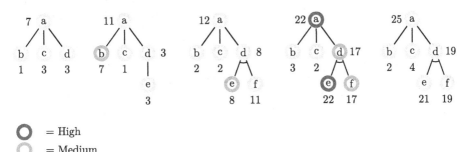

Fig. 1. An attack-tree series. The numbers indicate attribute γ and colours indicate temperature function τ. (Color figure online)

3.2 Attributes

An attribute is an abstract notion expressing a property of interest concerning the attack tree or its nodes, such as *cost of attack* [30]. Given an attack-tree series $(t_\delta)_{\delta \in \Delta}$, an attribute is represented by a series of valuation functions $\alpha_\delta: labels(t_\delta) \to D_\alpha$ for each time point $\delta \in \Delta$. The set D_α determines the range of values that can be taken by the attribute. We don't require that a valuation is a total function, thus allowing for valuations defined on a subset of a tree's nodes, such as the leaves.

There are various ways to determine the values of an attribute. Attributes can, for instance, be determined through observing events or by expert's opinions. Alternatively, an attribute can be completely determined by other attributes at the same time point. We call such an attribute a *derived* attribute.

For instance, if we have an attribute *direct damage* and an attribute *reputation damage*, then the sum of these two defines the derived attribute *total damage*.

If, for any time point δ_0, an attribute depends on the values of other attributes for time points $\delta \leq \delta_0$ then we say that this attribute is *history-dependent*. By restricting this to a fixed prefix of δ_0 with size n, we define the subclass of *n-history-dependent* attributes. The class of derived attributes then corresponds to 0-history-dependent attributes.

For the temporal analysis of an attack-tree series we introduce the notion of *temperature*. Let Γ be a finite and totally ordered set, called the *temperature domain*. A *temperature* is a history-dependent attribute with range Γ. The intuition behind this notion is that nodes with a high temperature indicate that they are of interest to an analyst observing the development of the attack tree over time. Consider, for instance, an attribute γ counting the *observed number of occurrences of an attack*, then the difference between the current and previous value of γ could be used to highlight an increasing prevalence of a certain attack type. A relevant temperature in this case would be the 1-history dependent attribute τ defined by $\tau_\delta(\ell) = \gamma_\delta(\ell) - \gamma_{\delta-1}(\ell)$.

Following these definitions, we can now define the challenge of a time-dependent analysis of a developing attack tree. Given an attack-tree series $(t_\delta)_{\delta \in \Delta}$, the challenge is to design one or more temperature attributes that relate to relevant security aspects of the system modeled by the attack-tree sequence, allowing the analyst to quickly observe security-related developments.

For an example of an attribute and a temperature function, we consider again Fig. 1. The attribute γ is defined through the values attributed to the nodes, e.g. $\gamma_3(a) = 12$. This particular attribute satisfies the property that conjunctive refinement is interpreted by the *min* function, while disjunctive refinement is interpreted by the *sum* function, e.g. $\gamma_3(a) = \gamma_3(b) + \gamma_3(c) + \gamma_3(d) = 2 + 2 + 8 = 12$.

We use colours of varying intensity to display the temperature values. The temperature attribute $\tau_\delta \colon \{a, b, c, d, e, f\} \to \{High, Medium, Low\}$ is defined by

$$\tau_\delta(\ell) = \begin{cases} High & \text{if } \gamma_{\delta-1} \text{ is defined and } \gamma_\delta(\ell) - \gamma_{\delta-1}(\ell) \in [10, \infty), \\ Medium & \text{if } \gamma_{\delta-1} \text{ is defined and } \gamma_\delta(\ell) - \gamma_{\delta-1}(\ell) \in [5, 10), \\ Low & \text{otherwise.} \end{cases}$$

The same attack-tree series with a different temperature attribute is shown in Fig. 2. The temperature attribute used in this case is 0-history dependent. It simply splits up the range of γ into four intervals. The advantage of the previously defined temperature function τ over τ' is that it only triggers if there is a sudden increase of γ.

$$\tau'_\delta(\ell) = \begin{cases} Very\ high & \text{if } \gamma_\delta(\ell) \in [15, \infty), \\ High & \text{if } \gamma_\delta(\ell) \in [10, 15), \\ Low & \text{if } \gamma_\delta(\ell) \in [5, 10), \\ Very\ low & \text{otherwise.} \end{cases}$$

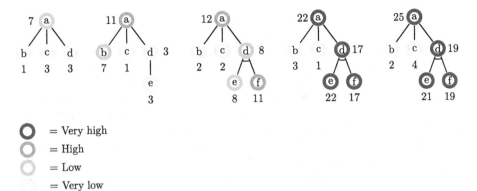

Fig. 2. An attack-tree series with the same attribute γ and a different temperature function τ'.

4 Usage Scenarios

In this section we discuss application scenarios for attack-tree series.

4.1 Highlighting Historical Trends

The most straightforward usage scenario for the proposed model is to spot and highlight trends using historical data. Our approach allows analysts to consistently visualize changes in the data values on an evolving set of attack scenarios expressed in an attack-tree series. Particularly, the analyst may choose a *temperature* function that expresses how important a change in some data values is, by highlighting more drastic fluctuations. We demonstrate how this may be displayed in an intuitive manner using a small case study.

Case Study Description. We use part of the realistic attack tree from [6]. This attack tree captures different scenarios in automatic teller machines (ATMs). For our case study, we focus on the ATM fraud scenario. This part of the attack tree contains 20 nodes. For the sake of simplicity, we limit this case study to dynamic data values, and we do not consider evolution of the tree structure.

We assume a company performing risk assessment for the ATM fraud scenario is interested in historical trends for three attributes: *probability*, *impact* (monetary loss) and *risk*, where risk is a derived attribute computed as the product *probability* × *impact*. The company uses historical data to evaluate these attributes. Probability of an attack can be evaluated from the frequency of such attacks over a given period of time, e.g.:

$$\Pr(attack) \approx frequency(attack)/total\ number\ of\ ATMs$$

While frequency is an imperfect estimator, it expresses well historical trends, i.e., with an increased frequency of an attack, its probability also goes up, and vice-versa.

Impact of an attack is estimated as the maximal direct monetary loss stemming from this attack. To compute average impact over a year, the company sums the total amount of losses and then divides it by the number of attacks. Historical losses can vary over the years due to many factors, for example, the extent of an attack, purchased insurance, or a change in legislation.

The data for the ATM scenario can be acquired from industry-related catalogues. For example, the European Association for Secure Transactions (EAST) regularly shares with its members extended statistics on ATM attacks and incurred losses[1]. We have used historical data from the ATM Crime Report 2015, where we have found statistics on ATM fraud attacks in the period 2010–2015 in a selected European area. Since the attack tree we use is not fully mapped to the ATM Crime Report data, we only have statistics about frequencies and losses for 5 attack tree nodes out of 20: `ATM fraud`, `cash trapping`, `transaction reversal`, `card trapping`, and `card skimming`. We visualize historical trends on real data only for these nodes.

We show a simple 1-history dependent temperature function τ'' (where $om(x)$ denotes order of magnitude of x):

$$
\tau_\delta''(\ell) = \begin{cases}
High & \text{if } \gamma_{\delta-1} \text{ is defined and } \gamma_\delta(\ell) - \gamma_{\delta-1}(\ell) \in [3om(\gamma_{\delta-1}(\ell)), \infty), \\
Medium & \text{if } \gamma_{\delta-1} \text{ is defined and } \gamma_\delta(\ell) - \gamma_{\delta-1}(\ell) \in [om(\gamma_{\delta-1}(\ell)), 3om(\gamma_{\delta-1}(\ell))), \\
Low & \text{if } \gamma_{\delta-1} \text{ is undefined or } \gamma_\delta(\ell) - \gamma_{\delta-1}(\ell) \in [-om(\gamma_{\delta-1}(\ell)), om(\gamma_{\delta-1}(\ell))), \\
Very\ low & \text{if } \gamma_{\delta-1} \text{ is defined and } \gamma_\delta(\ell) - \gamma_{\delta-1}(\ell) \in [-\infty, -om(\gamma_{\delta-1}(\ell))).
\end{cases}
$$

This temperature function works for both probability and risk attributes. An animated demo of our visualisation approach is available online in our github repository[2].

Figure 3 shows a snapshot of our ATM fraud visualization with temperature function τ'' for the risk attribute in 2012–2013. It visualizes that the risk of `card skimming`, while being the most important contributor to the overall `ATM fraud` risk, did not change much from the previous year. Risk of `cash trapping` has noticeably decreased in 2013, while risk of `card trapping` has increased. Finally, risk of `transaction reversal` has increased drastically from the previous year.

Due to the lack of complete historical data, we make another visualization using synthetic random data assigned to leaf nodes. By applying the bottom-up computation approach we complete the decoration process and obtain data for all intermediary nodes. For each year, for the probability values we generate random values in $[0,1]$, and for the impact values we generate random integers in appropriate intervals (from $[0, 100]$ for losses from shoulder-surfing, to $[0, 10000]$ for losses from attacks involving damaging the ATM, such as installing skimmers and trappers). Then we apply the appropriate bottom-up computation rules for the probability and maximal cost to the defender attributes [16] to fully decorate the attack tree. Finally, we multiply each probability and impact data point to compute risk, and we visualize the trends on all attack tree nodes using the

[1] https://www.association-secure-transactions.eu/tag/atm-crime-report/.
[2] Visualizations and code are published at https://github.com/vilena/atreeseries_viz.

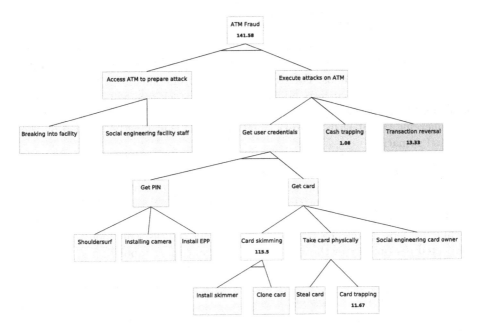

Fig. 3. A snapshot of the ATM fraud attack tree visualization on real data in 2013, with the temperature function τ'' for the *risk* attribute. The *high* value of the temperature function is shown in red, the *medium* value is shown in yellow, and the *very low* value is shown in green. The *low* values corresponding to very small changes in the attribute are shown as white nodes. We mark nodes for which we do not have data in grey. (Color figure online)

temperature function above. Animated visualizations with synthetic data are available online at our github repository.

The analyst looking at our visualization of historical trends for the ATM fraud scenario is able to identify the areas where the risk fluctuations are concentrated and review whether new countermeasures would be appropriate. However, this scenario also underlines the challenge to create informative temperature functions. Temperature function τ'' focuses on changes in a single node value. Thus, it may bring analyst attention to nodes where individual values change significantly, but they do not correspond to a major contribution to the value at the root node. In our example with the ATM Crime Report data, risk values at the `transaction reversal` node change considerably, due to the fact that it is a rare attack and it does not appear in the logs every year. Still, this attack is also not expensive for the ATM owner, as the risk value shows. The analyst may not want to investigate this type of attack, as it does not contribute significantly to the overall risk. Therefore, they may want to design a different temperature function, e.g., commesurable to the overall attribute value at the root node.

4.2 Other Application Scenarios

Spotting Emerging Trends. The visualization approach presented in the previous subsection can be integrated in a security dashboard to enable quick identification of emerging trends. For example, if the probability of a certain attack scenario goes up, the corresponding area in the attack tree will be highlighted, commanding attention of the security manager. This notification will highlight the emerging threat in its context (related attacks that may be enabled by the affected scenario).

Forecasting and Trend Extrapolation. Time-series data analysis is widely used to forecast future trends and extrapolate the ongoing dynamics [4]. Thus, our attack tree model naturally supports these usage scenarios for suitable data attributes. For instance, an organization may want to forecast, with regression techniques, how frequencies of social engineering attacks will change given the projected personnel growth rate. It may also want to extrapolate frequencies of probing attacks or cost of certain attack steps (e.g., cost of exploits or number of vulnerabilities in a software suite) given the dynamics observed locally or acquired from threat intelligence feeds.

What-if Analysis. The proposed attack tree model lends itself very well to what-if analysis. For instance, the analyst may evaluate potential consequences of a risk management decision to avoid or accept some attack scenarios (by removing or, respectively, adding/keeping some branches of the tree). Our consistency notion allows to investigate these planned scenarios and perform quantitative analysis in a coherent manner.

The analyst may also consider advanced scenarios when many data values are modified simultaneously or in sequence. For example, they can simulate several possible attack data dynamics (cost of certain attack steps goes down or stays the same), evaluating how resilient the company is to adverse event developments.

Security Investment Analysis. Last but not least, one of the main goals of security risk assessment and threat modelling is to identify missing security controls and prioritize security investments. The *temperature* function applied to attack-tree series visualization focuses the analyst's attention on critical attack steps or areas of the attack tree that are affected by the data dynamics. The analyst may elaborate from this visualization the right abstraction or system level where defences need to be positioned. They may also modify the data values in the what-if analysis fashion to investigate whether reducing probability or affecting cost of an attack via positioning a security control strategically would reduce the risk to the desired level. Additionally, if a company wants to purchase cyber-insurance, defender's costs may be projected considering the envisaged premiums that will depend on the company size and assets involved.

5 Conclusions

In this paper we have developed the notion of attack-tree series that enable modeling, visualization and analysis of dynamic scenarios with attack trees. We have also defined the notion of temperature functions as a special type of quantitative attributes that capture the importance of changes in the data values. We have visualized these ideas on a case study with the ATM fraud attack tree.

This preliminary work extends the theory of attack trees towards the demands of security data analysis and visualization. We plan to extend it in several directions. Firstly, we will further develop the theory of attack-tree series and temperature functions for attack-defense trees, thereby explicitly integrating security controls and facilitating security investment analysis.

In our current development, a temperature function is a history-dependent attribute, meaning that it is based on the history of values of an attribute. This works well if the snapshots are taken at regular intervals. However, if the time elapsed between the various snapshots of an attack-tree series is not constant, the temperature function may give a distorted result. Therefore, a straightforward generalisation would be to allow the temperature function to not only depend on the previous attack trees, but also on their time points.

As exemplified by our case study on real data, data values may not be available for some leaf nodes. In such cases, the bottom-up computation technique is infeasible. Buldas et al. [5] have recently shown that intermediary data values can be used to complete the attack tree decoration process. Our current theory of attack-tree series with attributes, including temperature functions, is agnostic to the decoration process, as it only requires that all attack trees in the time-series are decorated. We plan to further develop a more general theory for quantitative problems on attack-tree series that will take into account data decoration algorithms and the types of data available in threat intelligence feeds (e.g., the number of infections).

Finally, we would like to develop a system for automatic attack-tree series design and visualization from threat intelligence feeds. Recent work [10,13,14,26,34] demonstrated the viability of generating attack trees automatically. Particularly, Jhawar et al. [14] have shown that a threat library can be used to compose attack trees. A well-defined attack library like MITRE ATT&CK[3] could be used to automatically annotate information from threat feeds and add new relevant attack scenarios to an existing attack-tree series.

References

1. Amenaza. Securitree software (2017)
2. Amoroso, E.G.: Fundamentals of Computer Security Technology. Prentice-Hall Inc., Upper Saddle River (1994)
3. Bagnato, A., Kordy, B., Meland, P.H., Schweitzer, P.: Attribute decoration of attack-defense trees. Int. J. Secure Softw. Eng. 3(2), 1–35 (2012)
4. Box, G.E.P., Jenkins, G.M., Reinsel, G.C., Ljung, G.M.: Time Series Analysis: Forecasting and Control. Wiley, Hoboken (2015)
5. Buldas, A., Gadyatskaya, O., Lenin, A., Mauw, S., Trujillo-Rasua, R.: Attribute evaluation on attack trees with incomplete information. Computers & Security (2019, to appear)

---

[3] https://attack.mitre.org/.

6. Fraile, M., Ford, M., Gadyatskaya, O., Kumar, R., Stoelinga, M., Trujillo-Rasua, R.: Using attack-defense trees to analyze threats and countermeasures in an ATM: a case study. In: Horkoff, J., Jeusfeld, M.A., Persson, A. (eds.) PoEM 2016. LNBIP, vol. 267, pp. 326–334. Springer, Cham (2016). https://doi.org/10.1007/978-3-319-48393-1_24

7. Gadyatskaya, O., Hansen, R.R., Larsen, K.G., Legay, A., Olesen, M.C., Poulsen, D.B.: Modelling attack-defense trees using timed automata. In: Fränzle, M., Markey, N. (eds.) FORMATS 2016. LNCS, vol. 9884, pp. 35–50. Springer, Cham (2016). https://doi.org/10.1007/978-3-319-44878-7_3

8. Gadyatskaya, O., Harpes, C., Mauw, S., Muller, C., Muller, S.: Bridging two worlds: reconciling practical risk assessment methodologies with theory of attack trees. In: Kordy, B., Ekstedt, M., Kim, D.S. (eds.) GraMSec 2016. LNCS, vol. 9987, pp. 80–93. Springer, Cham (2016). https://doi.org/10.1007/978-3-319-46263-9_5

9. Gadyatskaya, O., Jhawar, R., Kordy, P., Lounis, K., Mauw, S., Trujillo-Rasua, R.: Attack trees for practical security assessment: ranking of attack scenarios with ADTool 2.0. In: Agha, G., Van Houdt, B. (eds.) QEST 2016. LNCS, vol. 9826, pp. 159–162. Springer, Cham (2016). https://doi.org/10.1007/978-3-319-43425-4_10

10. Gadyatskaya, O., Jhawar, R., Mauw, S., Trujillo-Rasua, R., Willemse, T.A.C.: Refinement-aware generation of attack trees. In: Livraga, G., Mitchell, C. (eds.) STM 2017. LNCS, vol. 10547, pp. 164–179. Springer, Cham (2017). https://doi.org/10.1007/978-3-319-68063-7_11

11. Garae, J., Ko, R.K.L.: Visualization and data provenance trends in decision support for cybersecurity. In: Palomares Carrascosa, I., Kalutarage, H.K., Huang, Y. (eds.) Data Analytics and Decision Support for Cybersecurity. DA, pp. 243–270. Springer, Cham (2017). https://doi.org/10.1007/978-3-319-59439-2_9

12. Green, I.: Extreme cyber scenario planning & attack tree analysis (2013). Talk at RSA Conference https://www.rsaconference.com/writable/presentations/file_upload/grc-t17.pdf

13. Ivanova, M.G., Probst, C.W., Hansen, R.R., Kammüller, F.: Attack tree generation by policy invalidation. In: Akram, R.N., Jajodia, S. (eds.) WISTP 2015. LNCS, vol. 9311, pp. 249–259. Springer, Cham (2015). https://doi.org/10.1007/978-3-319-24018-3_16

14. Jhawar, R., Lounis, K., Mauw, S., Ramírez-Cruz, Y.: Semi-automatically augmenting attack trees using an annotated attack tree library. In: Katsikas, S.K., Alcaraz, C. (eds.) STM 2018. LNCS, vol. 11091, pp. 85–101. Springer, Cham (2018). https://doi.org/10.1007/978-3-030-01141-3_6

15. Kordy, B., Mauw, S., Radomirovic, S., Schweitzer, P.: Attack-defense trees. J. Logic Comput. 24(1), 55–87 (2014)

16. Kordy, B., Mauw, S., Schweitzer, P.: Quantitative questions on attack–defense trees. In: Kwon, T., Lee, M.-K., Kwon, D. (eds.) ICISC 2012. LNCS, vol. 7839, pp. 49–64. Springer, Heidelberg (2013). https://doi.org/10.1007/978-3-642-37682-5_5

17. Kordy, B., Kordy, P., Mauw, S., Schweitzer, P.: ADTool: security analysis with attack–defense trees. In: Joshi, K., Siegle, M., Stoelinga, M., D'Argenio, P.R. (eds.) QEST 2013. LNCS, vol. 8054, pp. 173–176. Springer, Heidelberg (2013). https://doi.org/10.1007/978-3-642-40196-1_15

18. Kumar, R., Stoelinga, M.: Quantitative security and safety analysis with attack-fault trees. In: Proceedings 18th International Symposium on High Assurance Systems Engineering (HASE 2017), pp. 25–32. IEEE (2017)

19. Lakkaraju, K., Yurcik, W., Lee, A.J.: NVisionIP: netflow visualizations of system state for security situational awareness. In: Proceedings 2004 ACM Workshop on Visualization and Data Mining for Computer Security (VizSEC/DMSEC 2004), pp. 65–72. ACM (2004)

20. Li, E., Barendse, J., Brodbeck, F., Tanner, A.: From A to Z: developing a visual vocabulary for information security threat visualisation. In: Kordy, B., Ekstedt, M., Kim, D.S. (eds.) GraMSec 2016. LNCS, vol. 9987, pp. 102–118. Springer, Cham (2016). https://doi.org/10.1007/978-3-319-46263-9_7

21. Liao, X., Yuan, K., Wang, X.F., Li, Z., Xing, L., Beyah, R.: Acing the IOC game: toward automatic discovery and analysis of open-source cyber threat intelligence. In: Proceedings of the 2016 ACM SIGSAC Conference on Computer and Communications Security, pp. 755–766. ACM (2016)

22. Mauw, S., Oostdijk, M.: Foundations of attack trees. In: Won, D.H., Kim, S. (eds.) ICISC 2005. LNCS, vol. 3935, pp. 186–198. Springer, Heidelberg (2006). https://doi.org/10.1007/11734727_17

23. Noel, S., Harley, E., Tam, K.H., Limiero, M., Share, M.:. CyGraph: graph-based analytics and visualization for cybersecurity. In: Handbook of Statistics, vol. 35, pp. 117–167. Elsevier (2016)

24. Paul, S.: Towards automating the construction & maintenance of attack trees: a feasibility study. In: Proceedings 1st International Workshop on Graphical Models for Security (GraMSec 2014), Grenoble, France, volume 148 of EPTCS, pp. 31–46 (2014)

25. Paul, S., Vignon-Davillier, R.: Unifying traditional risk assessment approaches with attack trees. J. Inf. Secur. Appl. **19**(3), 165–181 (2014)

26. Pinchinat, S., Acher, M., Vojtisek, D.: ATSyRa: an integrated environment for synthesizing attack trees. In: Mauw, S., Kordy, B., Jajodia, S. (eds.) GraMSec 2015. LNCS, vol. 9390, pp. 97–101. Springer, Cham (2016). https://doi.org/10.1007/978-3-319-29968-6_7

27. Rasmussen, J., Ehrlich, K., Ross, S., Kirk, S., Gruen, D., Patterson, J.: Nimble cybersecurity incident management through visualization and defensible recommendations. In: Proceedings 7th International Symposium on Visualization for Cyber Security (VizSec 2010), pp. 102–113. ACM (2010)

28. Roy, A., Kim, D.S., Trivedi, K.S.: Attack countermeasure trees (ACT): towards unifying the constructs of attack and defense trees. Secur. Commun. Netw. **5**(8), 929–943 (2012)

29. Salter, C., Saydjari, O.S., Schneier, B., Wallner, J.: Toward a secure system engineering methodology. In: Proceedings 1998 Workshop on New Security Paradigms (NSPW 1998), pp. 2–10. ACM (1998)

30. Schneier, B.: Attack trees: modeling security threats. Dobb's J. Softw. Tools **24**(12), 21–29 (1999)

31. Shostack, A.: Threat Modeling: Designing for Security. Wiley, Hoboken (2014)

32. Takahashi, T., Emura, K., Kanaoka, A., Matsuo, S., Minowa, T.: Risk visualization and alerting system: architecture and proof-of-concept implementation. In: Proceedings 1st International Workshop on Security in Embedded Systems and Smartphones (SESP 2013), pp. 3–10. ACM (2013)

33. Tounsi, W., Rais, H.: A survey on technical threat intelligence in the age of sophisticated cyber attacks. Comput. Secur. **72**, 212–233 (2018)

34. Vigo, R., Nielson, F., Nielson, H.R.: Automated generation of attack trees. In: Proceedings 27th IEEE Computer Security Foundations Symposium (CSF 2014), pp. 337–350. IEEE (2014)

Attack Trees

Attack Trees: A Notion of Missing Attacks

Sophie Pinchinat[1(✉)], Barbara Fila[2], Florence Wacheux[1],
and Yann Thierry-Mieg[3]

[1] Univ Rennes, CNRS, IRISA, Rennes, France
`sophie.pinchinat@irisa.fr`
[2] Univ Rennes, INSA Rennes, CNRS, IRISA, Rennes, France
[3] Sorbonne Université, CNRS, LIP6, Paris, France

Abstract. Attack trees are widely used for security modeling and risk
analysis. Classically, an attack tree combines possible actions of the
attacker into attacks. In most existing approaches, an attack tree represents generic ways of attacking a system, but without taking any specific
system or its configuration into account. This means that such a generic
attack tree may contain attacks that are not applicable to the analyzed
system, and also that a given system could enable some attacks that the
attack tree did not capture.

To overcome this problem, we extend the attack tree setting with
a model of the analyzed system, allowing us to introduce precise *path
semantics* of an attack tree and to define *missing attacks*. We investigate
the *missing attack existence problem* and show how to solve it by calls
to the NP oracle that answers the *trace attack tree membership problem*;
the latter problem has been implemented and is available as an open
source prototype.

Keywords: Risk analysis · Attack trees · Path semantics · Missing
attacks · Complexity

1 Introduction

Attack trees are well-known graphical models for security analysis. They were
introduced by Schneier twenty years ago [23] and since then they have been successfully adopted by industry [1,7,8,11,19], as well as gained a lot of popularity
within the scientific security community [10,15]. The hierarchical structure of
an attack tree refines the goal of an attacker (depicted by the root node) into
simpler sub-goals, using disjunctive and conjunctive nodes. At the bottom of
the tree we find the leaves that correspond to actions that the attacker needs to
perform to reach his goal. Combinations of such actions form attacks that lead
to achieving the root goal of the tree. Numerous formalizations of attacks in the
context of attack trees have been proposed: multisets [18], sets [16], models of
Boolean functions [14], special types of graphs [13], paths or traces [2,3], etc. In
this paper, we model possible attacks as sequences of the attacker's actions.

© Springer Nature Switzerland AG 2019
M. Albanese et al. (Eds.): GraMSec 2019, LNCS 11720, pp. 23–49, 2019.
https://doi.org/10.1007/978-3-030-36537-0_3

A lot of research effort has recently been put into the problem of generation of attack trees [3,9,12,21,25]. Regardless of whether such generation is manual or automated, two main approaches can be distinguished: a *generic approach*, where the constructed attack tree covers a large number of well-known, generic attacks, applicable to most potential attackers; and a *specific approach*, where an attack tree is tailored to a specific system and/or to a specific attacker profile. In the first case, the creation process may benefit from the usage of attack tree libraries or commonly-known attack patterns, such as [19], whereas in the second case, a suitable attack tree is usually derived from a formal model of the analyzed system, e.g., [12,20,21]. Unfortunately, both approaches have their limitations. On the one hand, attack trees created using the generic approach may contain attacks that are irrelevant for the analysis of a given system – that we call *extra attacks*. On the other hand, in the case of the specific approach, the modeler can easily miss some attacks, for instance because he is not aware of particularities of the system. In this case, we talk about *missing attacks*. In both – generic and specific – approaches, it is also possible that the tree contains some weird attacks or misses some other ones due to an inappropriate expertise of the modeler, or because the formal model of the analyzed system is too coarse or too abstract.

We argue that, in order to perform a decent security analysis, an attack tree model needs to be coupled with the formal model of the analyzed system. Indeed, the former represents how the system can be attacked, whereas the latter describes how this system actually looks like. Taking both models into account simultaneously provides an elegant way of formally verifying the relevance of an attack tree w.r.t. the system, in terms of extra and missing attacks. The presence of the system model also allows us to extend the attack tree formalism with *weak* refinement operators that are used to refine goals in a more flexible manner. The specific contributions of this work are the following:

- We accompany an attack tree with an explicit modeling of the analyzed system, using a labeled transition system, which allows us to propose a new semantics for attack trees – *path semantics* – formalizing attacks in terms of sequences of attacker's actions corresponding to paths in the analyzed system. The use of this model of the system automatically discards extra attacks in the semantics.
- We extend OR-AND-SAND attack trees with the weak conjunctive (wAND) and weak sequential (wSAND) operators, allowing an expert to model collections of actions that are necessary but might not be sufficient for the attacker to reach his goal.
- We formally define two decision problems: *trace attack tree membership* (TATM) and *missing attack existence* (MAE). The first one focuses on whether a given attack is covered by an attack tree; the second, whether the tree contains any missing attacks w.r.t. the considered system.
- We provide algorithms solving the two problems. We prove that TATM is NP-complete, and that MAE for trees with no weak operators is in the second level of the polynomial hierarchy [24] resorting to the NP oracle for TATM.

This paper is structured as follows. Section 2 describes the relevant existing work. In Sect. 3, we present the background knowledge on attack trees that is necessary to understand the framework proposed in this article. We extend the attack tree model with a formalization of the system under study in Sect. 4. We introduce the concept of missing attacks, study the missing attack existence decision problem, and keep on with the trace membership problem in Sect. 5. In this section we also briefly describe the tool that we implemented to automate the solving of the TATM problem. We conclude and discuss future research directions in Sect. 6.

2 Related Work

Although the attack tree literature is very abundant [10,15], only two lines of research involve an explicit modeling of the analyzed system.

The first direction of work combining attack trees and a system model concentrates on attack tree correctness. In [2] and [4], a transition system whose states are labeled with propositions that express possible configurations of the underlying real-life system is employed. The authors define a novel kind of attack trees, called *state-based attack trees*, where the attacker's goals, i.e., nodes' labels, are expressed with two propositions representing the initial configuration, from which the attacker starts his attack, and a final configuration that the attacker aims at. Since the system model and the attack tree use a common language of propositions, it is possible to identify the set of paths in the system that allow the attacker to reach the goal of a specific node, i.e., to go from its initial to its final configuration. By using appropriate combination operators, similar to the sequential and parallel composition of paths used it this work, one can thus check whether an attack tree represents at least one valid attack in the analyzed system (non-emptiness problem [4]), as well as verify the quality of a node's refinement w.r.t. the system [2], i.e., check whether all paths satisfying the goal of the parent node also satisfy the combination of the goals of its children (over-match), and vice-versa (under-match).

The concept of path semantics used in the current work is very similar to the one from [2] and [4]. The main difference consists in the way in which attack tree node labels are formalized: goals over propositions, in the state-based approach, versus attacker's actions and their combinations, in the present work.

Supporting attack tree generation is the second research direction that involves reasoning about a particular system in the context of attack trees. The objective of [12] is to create an attack tree describing how a given socio-technical system, e.g., a company, can be attacked. The system is represented with a graph-based model capturing its locations, actors and processes involved, and relevant assets. Conditions, called policies, define which actions can be performed by which actor or process, and how. An attacker's goal is expressed in terms of policy invalidation: actors, assets, and locations necessary to enact a policy are determined, and the corresponding path in the system model is identified. An algorithm recursively constructs an attack tree for every policy, and then combines them using an AND node to get a tree representing a single path. Finally, the trees corresponding to particular paths are combined under a common OR node, resulting in a tree invalidating the initial policy.

In [20] and [21], the authors address the problem of generating an attack tree for a physical system formalized using a domain specific language. The system description is compiled into a symbolic transition system, and the reachability analysis, based on model checking, is performed to generate attack scenarios expressed as sequences of elementary actions of the attacker. By combining these conjunctive scenarios using an OR node, a flat attack tree is obtained. Parsing and merging are then used to factorize the tree and obtain a more usable and efficient representation.

In [25], a system is modeled with the help of a particular type of process algebra, called value-passing quality calculus. Given a target location or an asset of interest in the system, an AND-OR attack tree representing how the attacker may reach the location or acquire the asset is constructed using SAT solving.

An explicit modeling of the system has also been exploited in [3], where a state-based attack tree is incrementally derived from a quantitative analysis of the transition system representing the real-life system to be analyzed. First, optimal paths, e.g., those corresponding to the cheapest or the fastest attacks, are determined in the transition system, and they are then used to identify leaves that contribute to these optimal attacks and could therefore be interesting candidates for further refinement.

The work described in [9] uses a similar system model as [3]. The states of the transition system are represented by the set of predicates valid in this state. The objective is to generate an attack tree based on a set of successful traces in the transition system, i.e., traces that start from the initial state and end in any state containing a desired set of predicates. The particularity of the obtained tree is that it is refinement-aware, i.e., that its nodes correspond to the meaningful levels of abstraction that can be expressed using the underlying transition system components. It is to be noted that the resulting tree contains only OR and SAND refinements, i.e., no classical AND operator is used. In the worst case, this may imply that the size of the produced tree is exponentially larger than if the AND refinement was used.

In all generation approaches described above, attack trees are synthesized for a given system from its formal model. They thus capture only the attacks that apply to this specific system. In contrast, in our framework, an attack tree might be generic, and it is put in context of an, *a priori*, independent system. Our goal is to determine which of the attacks covered by the tree are indeed applicable to the given system, and which of the paths in the system correspond to attacks in practice, but are not covered by the tree. To the best of our knowledge, the problem of missing attacks has not yet been formally investigated.

A second novelty of our work is a formalization of *weak* refinement operators for attack trees. The weak operators capture the fact that some actions, although not explicitly present in a tree, might be required so that the path in the system is indeed an attack. It turns out that a very similar issue has recently been studied by Mantel and Probst in [17], where the authors introduced the *purity* property. This property stipulates whether an attack should perfectly fit a sequence of an attack tree leaves or whether they can be interleaved with other actions. The core problem addressed by our weak refinement operators and by the purity property is the same, but some difference can be observed. Our weak operators are used locally at a given node, so it is possible to accept some additional actions at some but not at all nodes of the tree. In contrast, the purity criterion of [17] seems to be defined as a property of the attack tree semantics, so the additional actions are allowed either at all nodes or at none.

3   Background on Attack Trees

We start with an explanation of what an attack tree is, in Sect. 3.1, before introducing the notion of an attack in Sect. 3.2.

3.1   Attack Trees Informally

Intuitively speaking, an attack tree is a labeled tree representing how an attacker can proceed to attack a system. The label of the root node describes the main goal of the attacker and the remaining nodes refine this goal into subgoals. In this work, we use classical refinement operators – *disjunctive* (OR), *conjunctive* (AND), and *sequential* (SAND) – as well as new, weak operators – *weak conjunctive* (wAND) and *weak sequential* (wSAND).

The attack tree leaves represent goals that are precise enough and thus do not need to be refined any further. The goal of an OR node is achieved if at least one of the goals of its children is achieved. Achievement of the goal of an AND node requires to achieve the goals of all of its children. The goal of a SAND node is achieved if the goals of all of its children are achieved in the specified order. To achieve the goal of a node refined using wAND (resp. wSAND), the attacker needs to achieve the goals of all of its children (resp. in the given order) but in addition, some other actions (not necessarily under the control of the attacker) may also be necessary before the goal of the node can be fully reached. For instance, consider that to be able to attack a system, the attacker needs to deactivate several alarms. In order to do so, he shuts down the power supply, and pursues his attack. However, after a short period of time, the back-up power supply takes over and the alarms are back on. The action of putting the electricity back on is not the attacker's action, so it will not appear in an attack tree explicitly. To make modeling of such attack scenarios possible, we use the weak refinement operators.

Example 1. An example of an attack tree is given in Fig. 1, where we use standard notation: arcs denote conjunctive nodes, and arrows sequential ones. We will later use dotted arcs and arrows for weak operators. The main goal of the attacker is to get a document. To achieve this, the attacker may either corrupt an employee, or steal the document by himself. To corrupt the employee, the attacker may bribe or blackmail him. Stealing the document requires penetrating the building and taking the document. To enter the building, the attacker must unlock the door and then enter undetected. The lock can be opened with a key that would need to be stolen or it can be forced. The attacker enters undetected if he manages to deactivate the alarm and pass the door.

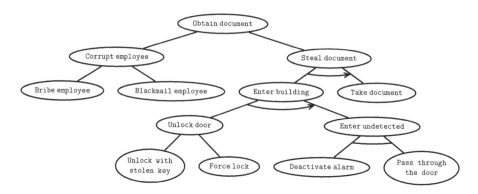

Fig. 1. An attack tree for obtaining a document.

We now give the definition and formal notation for attack trees. For the rest of the paper, we fix a set ACT that models all the actions that the attacker can execute.

Definition 1. *An* attack tree θ *over* ACT *is either a leaf* $a \in$ ACT, *or a composed tree* OP$(\theta_1, \theta_2, \ldots, \theta_n)$, *where* OP $\in \{$ OR, SAND, AND, wSAND, wAND$\}$ *and* $\theta_1, \theta_2, \ldots, \theta_n$ *are attack trees over* ACT.

To differentiate classical OR, AND, and SAND refinement operators from wAND and wSAND, we refer to the former ones as *strong* and to the latter ones as *weak*.

3.2   Attacks

An attack tree represents a collection of attacks that an attacker can follow to achieve his goal. In this work we formalize an *attack* as a sequence of the elements of ACT sufficient to achieve the goal of the root node of the tree, accordingly to the refinement operators.

Example 2. For instance, the tree in Fig. 1 models the following set of attacks:

A_1 :   Bribe employee

A_2 :   Blackmail employee

A_3 :   Unlock with stolen key, Deactivate alarm, Pass through the door, Take document

A_4 :   Unlock with stolen key, Pass through the door, Deactivate alarm, Take document

A_5 :   Force lock, Deactivate alarm, Pass through the door, Take document

A_6 :   Force lock, Pass through the door, Deactivate alarm, Take document

One can notice that attacks A_3 and A_4 (similarly A_5 and A_6) differ only in the order in which the actions of passing through the door and deactivating the alarm are executed. This is due to the AND operator refining the Enter undetected node. However, given a specific building where the document is stored, one of these attacks might actually be unfeasible. Indeed, some alarms can be deactivated only from outside, whereas others require a person to first enter the building and then deactivate the alarm within a short, predefined time lapse. Since attack trees are often created without having full knowledge of the system they will be applied to, they might represent some sequences of actions that in reality are not valid attacks. On the contrary, the attack tree modeler who happens to have a specific type of alarm at his home, e.g., the one to be deactivated from outside, might be biased during the attack tree creation process and model Enter undetected with SAND (Deactivate alarm, Pass through the door), in which case the sequences A_4 and A_6 will not be considered as possible attacks in the tree, even if they are feasible in the system.

In order to deal with such discrepancies, it is necessary to put an attack tree in the context of the analyzed system to define a formal semantics for attack trees that captures all valid attacks, and only them. We achieve this by an explicit modeling of the analyzed system, using labeled transition systems.

4   Enhancement of the Attack Tree Model

In Sect. 4.1, we recall basic knowledge on labeled transition systems and define operations on paths that we use in Sect. 4.2 to equip attack trees with formal semantics relative to the analyzed system.

4.1   System Modeling with Labeled Transition Systems

To model real-life systems whose security we want to analyze, we use transition systems with transitions labeled by the elements of ACT. We use non-deterministic systems to be able to capture the fact that some actions of the attacker are guided by the environment or are conditioned on the actions of other parties. Exploiting non-determinism to reason about an impact of the environment on an agent behavior is a standard approach in the model checking community (see for example [5, page 22]).

Definition 2. *A finite transition system labeled by* ACT *is a tuple* $\mathcal{S} \stackrel{def}{=} (S, \rightarrow)$, *where* S *is the set of states and* $\rightarrow \subseteq S \times$ ACT $\times S$ *is the set of transitions.*

Note that we write $s_1 \xrightarrow{a} s_2$ instead of $(s_1, a, s_2) \in \rightarrow$ when referring to a transition between two states $s_1, s_2 \in S$ labeled by $a \in$ ACT, and call this transition an a-transition.

An example of a very simple transition system, modeling how a person can deactivate two alarms, is given in Fig. 5, in Sect. 4.2.

Let us now recall the notion of a path in a transition system, that we use in our framework to formalize attacks.

Definition 3. *A* path π *in a transition system* $\mathcal{S}$ *is a finite sequence of the form* $\pi = s_0 a_1 s_1 \ldots a_n s_n$, *where, for all* $0 \leq i < n$, *we have* $s_i \xrightarrow{a_{i+1}} s_{i+1}$. *We let* $\pi.first \overset{def}{=} s_0$, $\pi.last \overset{def}{=} s_n$, *and* $trace(\pi) \overset{def}{=} a_1 \ldots a_n \in ACT^*$.

A path $\pi = s_0 a_1 s_1 \ldots a_n s_n$ *is* elementary *whenever the states occurring in* π *are all distinct, namely, for all* $i \neq j$, *we have* $s_i \neq s_j$.

The set of all paths (resp. elementary paths) in $\mathcal{S}$ is denoted by $\Pi(\mathcal{S})$ (resp. $\Pi_{elem}(\mathcal{S})$). Given a set of paths $\Pi \subseteq \Pi(\mathcal{S})$, we write $\Pi.first$ for the set $\{\pi.first \mid \pi \in \Pi\}$ and $\Pi.last$ for the set $\{\pi.last \mid \pi \in \Pi\}$.

The size of a path π, written $|\pi|$, is equal to its number of transitions, and $\pi(i)$ is the $i + 1^{st}$ state of π. We have $\pi(0) = \pi.first$ and $\pi(|\pi|) = \pi.last$. The subsequence of states in a path π from $\pi(i)$ to $\pi(j)$ is denoted by $\pi[i, j]$. Such a subsequence is called a *factor* of π with *anchoring* $[i, j]$.

In order to further analyze paths, we now define their concatenation and parallel composition. Intuitively speaking, the concatenation of two paths π_1 and π_2 can be done if the last state of π_1 is equal to the first state of π_2. The result of the concatenation is then a path π containing the sequence of states of π_1 followed by the sequence of states of π_2, without any additional state or transition, as illustrated in Fig. 2.

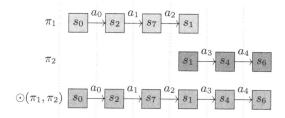

Fig. 2. Concatenation of paths π_1 and π_2.

Formally, the path concatenation is defined as follows:

Definition 4. *Let* $\pi_1, \pi_2, \ldots, \pi_n \in \Pi(\mathcal{S})$ *be paths in* $\mathcal{S}$, *such that* $\pi_i.last = \pi_{i+1}.first$, *for* $1 \leq i < n$. *The concatenation of* $\pi_1, \pi_2, \ldots, \pi_n$, *denoted with* $\odot(\pi_1, \pi_2, \ldots, \pi_n)$, *is the path* π *satisfying* $\pi[\sum_{k=1}^{i-1} |\pi_k|, (\sum_{k=1}^{i-1} |\pi_k|) + |\pi_i|] = \pi_i$, *for every* $i \in \{1, \ldots, n\}$.

By extension, given sets of paths $\Pi_1, \Pi_2, \ldots, \Pi_n$, *we let*

$$\odot(\Pi_1, \Pi_2, \ldots, \Pi_n) \overset{def}{=} \{\odot(\pi_1, \pi_2, \ldots, \pi_n) \mid \pi_i \in \Pi_i, \text{ for } 1 \leq i \leq n\}.$$

Concatenation on sets of paths will be used to define the semantics of the SAND operator in attack trees. To formalize the AND operator, we will use the notion of *parallel composition* of paths.

Intuitively speaking, a path π is a parallel composition of paths $\pi_1, \pi_2, \ldots, \pi_n$ if it is possible to obtain π by combining $\pi_1, \pi_2, \ldots, \pi_n$ in a way where every pair of consecutive states (i.e., every transition) in π is covered by one of the π_i. Thus, the very action carried by this transition takes part in at least one of the paths among $\pi_1, \pi_2, \ldots, \pi_n$. Figure 3 illustrates the notion of parallel composition of paths.

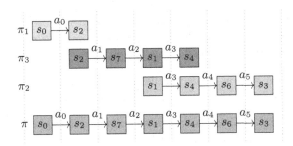

Fig. 3. Path π is a parallel composition of paths π_1, π_2, and π_3.

Figure 4 shows an example of a path π that is *not* a parallel composition of π_1, π_2, π_3, because the transition $s_2 \xrightarrow{a_1} s_7$ in π is not covered by any of the π_i's.

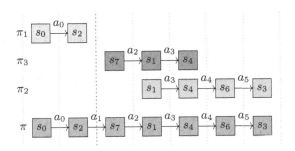

Fig. 4. Path π is not a parallel composition of π_1, π_2, π_3.

In Example 3 of Sect. 4.2, we expose the reasons for introducing such a parallel composition of paths, whose formal definition is as follows.

Definition 5. *A path π is a* parallel composition *of paths $\pi_1, \pi_2, \ldots, \pi_n$, denoted by $\pi \in \mathbb{M}(\pi_1, \pi_2, \ldots, \pi_n)$, whenever the following conditions are satisfied:*

- *for every $i \in \{1, \ldots, n\}$, the path π_i is a factor of π at some anchoring $[k_i, l_i]$,*
- *for every $j \in \{0, \ldots, |\pi| - 1\}$, the inclusion $[j, j+1] \subseteq [k_i, l_i]$ holds, for some $i \in \{1, \ldots, n\}$.*

By extension, given sets of paths $\Pi_1, \Pi_2, \ldots, \Pi_n$, we let

$$\mathbb{M}(\Pi_1, \Pi_2, \ldots, \Pi_n) \stackrel{def}{=} \{\mathbb{M}(\pi_1, \pi_2, \ldots, \pi_n) \mid \pi_i \in \Pi_i, \text{ for } 1 \leq i \leq n\}.$$

It is important to notice that our parallel composition of paths is orthogonal to the notion of parallel composition of labeled transition systems: parallel composition of paths defined in Definition 5 captures the concomitance of goal achievements along a fixed execution (path) of the transition system, while the parallel composition of transition systems reflects the concomitance of executions.

Finally, to formalize the weak refinement operators, we relax the parallel composition to its weak form, called *weak parallel composition*. In this case, the paths to be composed do not need to overlap at all.

Definition 6. *Path π is a* weak parallel composition *of paths $\pi_1, \pi_2, \ldots, \pi_n$, denoted by $\pi \in |||(\pi_1, \pi_2, \ldots, \pi_n)$, whenever the following holds:*

- *for every $i \in \{1, \ldots, n\}$, the path π_i is a factor of π at some anchoring $[k_i, l_i]$,*
- *$\pi[0, |\pi_i|] = \pi_i$, for some $i \in \{1, \ldots, n\}$,*
- *$\pi[|\pi| - |\pi_j|, |\pi|] = \pi_j$, for some $j \in \{1, \ldots, n\}$.*

By extension, given sets of paths $\Pi_1, \Pi_2, \ldots, \Pi_n$, we let

$$|||(\Pi_1, \Pi_2, \ldots, \Pi_n) \stackrel{def}{=} \{|||(\pi_1, \pi_2, \ldots, \pi_n) \mid \pi_i \in \Pi_i, \text{ for } 1 \leq i \leq n\}.$$

Note that in Fig. 4, while π is not a parallel composition of paths π_1, π_2, π_3, it is a weak parallel composition, i.e., $\pi \in |||(\pi_1, \pi_2, \pi_3)$.

In the next section, we use the operations on sets of paths defined here to construct the semantics of an attack tree in the presence of a system.

4.2   Attack Tree Semantics in the Presence of a System Model

Let $\mathcal{S}$ be a transition system labeled by ACT, and let θ be an attack tree over ACT. Our objective is to define the semantics of θ in terms of sequences of actions from $\mathcal{S}$, that satisfy the root goal of θ. Each node will thus be interpreted as a set of paths in $\mathcal{S}$, that is constructed as follows.

A leaf node labeled with $a \in$ ACT is simply interpreted with paths of length one, corresponding to a-transitions in $\mathcal{S}$. The interpretation of an OR node is the union of the sets of paths corresponding to its children. Indeed, any path satisfying the goal of a child of an OR node also satisfies the goal of the OR node itself.

To achieve the goal of a SAND node, the attacker needs to achieve the goals of all of its children in the given order. Thus, to provide the interpretation of a SAND node, we concatenate the sets of paths corresponding to its children, as defined in Definition 4. Similarly, to achieve the goal of a wSAND node, the goals of all of its children need to be achieved in the given order, but arbitrary other actions can occur between each subgoal realization. This permits to capture the interleaving of this sequential goal with other parts of the attack, as well as adequately model cases where some system reaction or behavior is needed to continue the attack (recall the example of the back-up power supply, discussed in Sect. 3.1). Formally, we thus interpret a wSAND node using concatenation of sets of paths satisfying the node's children interleaved with any path possible in the system.

Finally, AND and wAND require that the goals of all of their child nodes are achieved, but do not impose any order on this achievement. In the case of the strong operator AND, the actions of each subgoal must be seen contiguously, whereas for the weak operator wAND these subgoals can be arbitrarily interleaved with other actions. To capture this behavior, parallel (Definition 5) and weak parallel composition (Definition 6) of the sets of paths are used to provide the interpretation for the two types of conjunctively refined nodes.

Overall weak versions of the operators are more lenient than their strong counterparts. They introduce a weaker notion of precedence that fits the use of underspecified attack trees (some actions of the system are not precisely modeled), and more generally the fact that the immediate "Next" constraint of the strong operators is difficult to enforce in the context of a concurrent system. These weak operators let us express "stutter invariant" behavior with attack trees. In practice, for concurrent systems, stutter invariant property specification is often more relevant than using the full temporal logic with neXt. Definition 7 summarizes the above discussion.

Definition 7. *Given a transition system S, the* path semantics *of an attack tree θ is the set of paths $[\![\theta]\!]^S \subseteq \Pi(S)$, defined by induction as follows:*

- $[\![a]\!]^S = \{s_1 a s_2 \in \Pi(S) \mid s_1, s_2 \in S\}$,
- $[\![OR(\theta_1, \theta_2, \ldots, \theta_n)]\!]^S = \cup([\![\theta_1]\!]^S, [\![\theta_2]\!]^S, \ldots, [\![\theta_n]\!]^S)$,
- $[\![SAND(\theta_1, \theta_2, \ldots, \theta_n)]\!]^S = \odot([\![\theta_1]\!]^S, [\![\theta_2]\!]^S, \ldots, [\![\theta_n]\!]^S)$,
- $[\![AND(\theta_1, \theta_2, \ldots, \theta_n)]\!]^S = \wedge([\![\theta_1]\!]^S, [\![\theta_2]\!]^S, \ldots, [\![\theta_n]\!]^S)$,
- $[\![wSAND(\theta_1, \theta_2, \ldots, \theta_n)]\!]^S = \odot([\![\theta_1]\!]^S, \Pi(S), [\![\theta_2]\!]^S, \Pi(S), \ldots, \Pi(S), [\![\theta_n]\!]^S)$,
- $[\![wAND(\theta_1, \theta_2, \ldots, \theta_n)]\!]^S = |||([\![\theta_1]\!]^S, [\![\theta_2]\!]^S, \ldots, [\![\theta_n]\!]^S)$.

The example below illustrates the use of parallel composition of paths to interpret an AND node.

Example 3. Suppose that an attacker needs to deactivate two alarms. To do so, he can either disable Alarm 1 and Alarm 2 in any order, or simply shut down the power supply, which automatically deactivates both alarms. The transition system modeling these possibilities is given in Fig. 5. Suppose that the attacker's behavior is modeled with the attack tree from Fig. 6. The path semantics of this tree is composed of paths $s_0 s_1 s_2$, $s_0 s_3 s_2$, and $s_0 s_2$ (we omit the

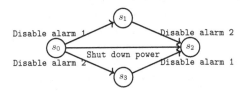

Fig. 5. Simple transition system.

actions for readability). The last path is interesting. Indeed, $s_0 s_2$ belongs to both $[\![\text{Deactivate Alarm 1}]\!]^S$ and $[\![\text{Deactivate Alarm 2}]\!]^S$. It is also a valid path in the parallel composition $\wedge([\![\text{DeactivateAlarm1}]\!]^S, [\![\text{DeactivateAlarm2}]\!]^S)$, because it satisfies both subgoals at the same time. This example shows that, in the parallel composition of paths, and thus in the semantics of AND nodes, any overlap between the paths interpreting their child nodes is allowed, as already illustrated in Fig. 3.

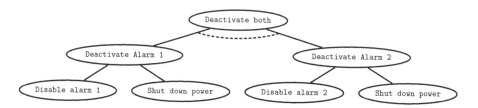

Fig. 6. Example with overlap on AND node.

Notice that the framework developed in this work allows for repetitions of an action in the tree, as it is the case of Shut down power in Example 3. The presence of such repeated actions may, for instance, result from the use of attack tree libraries. For some scenarios, having an action several times in a tree might be necessary. Recall the power example discussed in Sect. 3.1, where the back-up power supply is activated short after the main electricity source is switched off. Depending on how quickly the attacker performs his attack, the Shut down power action from the tree in Fig. 6 will need to be done either one or two times. To cover both cases, the action is repeated in the tree, and our formalism is flexible enough to interpret such repeated nodes as the same or separated instances of the action.

4.3 System-Based Approach to Classical View on Attack Trees

To finish this section, we relate the system-based view on attack trees with classical approaches where the system is not considered. Indeed, most of existing semantics for attack trees do not take the analyzed system into account. We would like to point out that our system-based framework can simulate such approaches by considering the *universal transition system over* ACT, written $\mathcal{U}_{\text{ACT}}$, allowing to execute any possible sequence of actions over ACT. The universal transition system is composed of a single state and a looping transition for each action in ACT, hence it looks like a flower. An example of $\mathcal{U}_{\text{ACT}}$ over ACT $= \{1, \dots, 9\}$ is given in Fig. 7.

Fig. 7. System $\mathcal{U}_{\{1,\dots,9\}}$.

To define the semantics of attack trees that is independent of the system, we introduce *trace language* of an attack tree θ over ACT, denoted by $\mathcal{L}(\theta) \subseteq \mathsf{ACT}^*$, and defined as:

$$\mathcal{L}(\theta) \overset{def}{=} \{trace(\rho) \mid \rho \in [\![\theta]\!]^{\mathcal{U}_{\mathsf{ACT}}}\}.$$

The following result makes a link between the trace language of an attack tree and its path semantics.

Proposition 1. *Given a transition system $\mathcal{S}$ and a path $\pi \in \Pi(\mathcal{S})$, we have that $\pi \in [\![\theta]\!]^{\mathcal{S}}$ if, and only if, $trace(\pi) \in \mathcal{L}(\theta)$.*

Proof. By definition of $\mathcal{L}(\theta)$, we have to show that $\pi \in [\![\theta]\!]^{\mathcal{S}}$ if, and only if, there exists $\rho \in [\![\theta]\!]^{\mathcal{U}_{\mathsf{ACT}}}$ with $trace(\pi) = trace(\rho)$. The candidate for ρ is the path like π but where each state is replaced by the unique state of $\mathcal{U}_{\mathsf{ACT}}$; note that ρ is indeed a path of $\mathcal{U}_{\mathsf{ACT}}$ and that by construction $trace(\pi) = trace(\rho)$. It is routine to verify by induction over θ that this candidate is adequate. $\square$

The following example emphasizes the role played by the system in the path semantics.

Example 4. Take the tree $\theta = \mathsf{SAND}(\mathsf{AND}(a, b), \mathsf{AND}(c, d))$. It is easy to see that $\mathcal{L}(\theta) = \{abcd, abdc, bacd, badc\}$. However, when interpreting the tree w.r.t. a system, using the semantics from Definition 7, not all sequences of actions will necessarily be feasible.

- If we consider the universal system $\mathcal{U}_{\mathsf{ACT}}$, then the path semantics will contain the four attacks.
- For the system $\mathcal{S}_1$ in Fig. 8a, the path semantics of the tree only contains attacks $abcd$ and $bacd$.
- Finally, for the system $\mathcal{S}_2$ in Fig. 8b, the path semantics contains attacks $abcd$, $abdc$, and $bacd$.

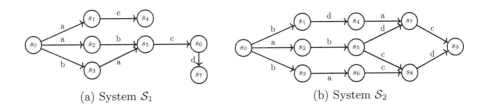

(a) System $\mathcal{S}_1$ (b) System $\mathcal{S}_2$

Fig. 8. Different systems induce different path semantics.

Example 4 illustrates that the same attack tree does not have to have the same meaning depending on the underlying system. The path semantics allows the experts to resort to some libraries of already designed trees without the inconvenience of checking what is indeed achievable or not in the system they have in mind. A practical example of such reusable attack tree libraries can be found in Chapter 4 of [19].

5   Missing Attacks

In this section, we exploit the path semantics to compare the tree with what the attacker can achieve in the system. In Sect. 5.1, we formally define missing attacks, and in Sects. 5.2, 5.3 and 5.4 we address the decision problem MAE for the existence of missing attacks together with the related trace membership decision problem TATM, and provide complexity bounds for these problems. Finally, Sect. 5.5 describes a prototype implementation that we developed to automate the solving of the TATM problem.

5.1   The Definition of Missing Attacks

We now explain how we can warn the security experts about possibly missing attacks in an attack tree. We start by an illustrating example. Consider the tree θ from Example 4 and let us look again at system $\mathcal{S}_2$ in Fig. 8b. What can we say about the sequence of actions $bdac$? According to system $\mathcal{S}_2$, this sequence starts and ends in the same states as actual attacks. Thus, the four paths between s_0 and s_9 are somehow equivalent in the system, but one of them is not in the semantics of attack tree θ. It could be interesting for the expert to get a warning about such paths that are equivalent to attacks in the system, but that he did not include in the tree. This is what we call *completeness analysis*.

Note that we are not claiming that any such path in the system *must* be considered as an attack, that is up to the experts to decide. However, we believe that analyzing the completeness of the attack tree and giving warnings to security experts about these paths can prevent some human-related errors, which can be troublesome if not tackled.

To formally define the notion of missing attack, we first introduce the closure of a set of paths that encompasses extra paths having the same extremities as the paths in the set.

Definition 8. *Given a system $\mathcal{S}$, the* closure *of a set of paths $\Pi \in \Pi(\mathcal{S})$ is* $cl(\Pi) \overset{def}{=} \{\pi \in \Pi_{elem}(\mathcal{S}) \mid \pi.first \in \Pi.first \text{ and } \pi.last \in \Pi.last\}$.

It is clear that any elementary path in Π is also in $cl(\Pi)$. However, the reciprocal may not hold in general and this is how we capture missing attacks.

Definition 9. *A path π is a* missing attack *if it belongs to $\Delta_\theta^{\mathcal{S}} \overset{def}{=} cl(\llbracket\theta\rrbracket^{\mathcal{S}})\setminus\llbracket\theta\rrbracket^{\mathcal{S}}$.*

To make it simple, a missing attack in θ is a path in $\mathcal{S}$ that imitates attacks in $\llbracket\theta\rrbracket^{\mathcal{S}}$ (starts and ends in states of existing attacks) but is not in $\llbracket\theta\rrbracket^{\mathcal{S}}$ itself. It is important to notice that missing attacks are restricted to elementary paths. This is a robust choice: indeed, if we allowed non-elementary paths, we would basically take an inventory of existing attacks extended with extra uninteresting cycles, irrelevant from the point of view of a rational attacker for adding useless sequences of actions.

Example 5. We illustrate the notion of missing attacks on the transition system from Fig. 5. Suppose that the tree designer is not interested in Alarm 2. We consider the leaf attack tree `Disable alarm 1`. We omit the actions for readability, and we have: $[\![\texttt{Disable alarm 1}]\!]^{\mathcal{S}} = \{s_0 \rightarrow s_1, s_3 \rightarrow s_2\}$, and $cl([\![\texttt{Disable alarm 1}]\!]^{\mathcal{S}}) = \{s_0 \rightarrow s_1, s_3 \rightarrow s_2, s_0 \rightarrow s_1 \rightarrow s_2, s_0 \rightarrow s_3 \rightarrow s_2, s_0 \rightarrow s_2\}$, so that $\Delta_\theta^{\mathcal{S}} = \{s_0 \rightarrow s_1 \rightarrow s_2, s_0 \rightarrow s_3 \rightarrow s_2, s_0 \rightarrow s_2\} \neq \emptyset$. In particular, the missing attack $s_0 \rightarrow s_2$ might be problematic if the expert protects the system against the action `Disable alarm 1` only, because there will still be a possibility for the attacker to counter the alarm by shutting down the power supply.

To allow for reasoning about missing attacks, we introduce and investigate the Missing Attack Existence decision problem.

5.2   The Missing Attack Existence Problem

Since in risk analysis missing attacks can have severe consequences, we address the natural question of the existence of missing attacks, captured by the following decision problem.

Definition 10 (Missing attack existence problem (MAE))

– *Input: an attack tree θ and a system $\mathcal{S}$.*
– *Output: $\Delta_\theta^{\mathcal{S}} \neq \emptyset$?*

Otherwise said, can we find three paths π, π_1, π_2 that satisfy the following constraints: $\pi.first = \pi_1.first$, $\pi.last = \pi_2.last$, $\pi \notin [\![\theta]\!]^{\mathcal{S}}$, $\pi_1 \in [\![\theta]\!]^{\mathcal{S}}$, and $\pi_2 \in [\![\theta]\!]^{\mathcal{S}}$?

It is very tempting to design a non-deterministic algorithm that can select three paths and then checks these five constraints above. However, due to potential weak operators wSAND and wAND, it seems difficult to bound the size of these paths. While the size of π can be bounded by the size of the system, as missing attacks are elementary paths, it is unclear how to bound the size of paths π_1 and π_2. Discarding weak operators gives a natural bound which is the number of leaves of θ, thus polynomial in the size of the input.

Now, once these three paths are guessed, the non-deterministic algorithm verifies the five constraints. The first two can be verified in $O(1)$, while the last three reduce to answering the *Trace Attack Tree Membership problem (TATM)* formalized in the following definition.

Definition 11 (Trace Attack Tree Membership problem (TATM))

– *Input: an attack tree θ (over ACT) and a trace $t \in ACT^*$*
– *Output: $t \in \mathcal{L}(\theta)$?*

Proposition 2. *TATM is* NP-*complete.*

Based on Proposition 2 (a corollary of Propositions 3 and 5 fully proven in the next Sects. 5.3 and 5.4 respectively), we design the non-deterministic Algorithm 1 that makes three independent calls to NP oracles.

Input: An attack tree θ, a system $\mathcal{S}$.
Output: ACCEPT if $\Delta_\theta^\mathcal{S} \neq \emptyset$.
1 **CHOOSE** elementary path $\pi \in \Pi(\mathcal{S})$, and $\pi_1, \pi_2 \in \Pi(\mathcal{S})$ of length at
   most $|\theta|$;
2 **if** $\pi.first \neq \pi_1.first$ or $\pi.last \neq \pi_2.last$ **then**
3 | **REJECT**
4 **end**
5 **else**
6 | **if** *the* NP *oracle for the question* "$trace(\pi) \in \mathcal{L}(\theta)$" *answers* "*Yes*"
   | **then**
7 | | **REJECT**
8 | **end**
9 | **else**
10 | | **if** *the* NP *oracle for the question* "$trace(\pi_1) \in \mathcal{L}(\theta)$" *answers*
   | | "*No*" **then**
11 | | | **REJECT**
12 | | **end**
13 | | **else**
14 | | | **if** *the* NP *oracle for the question* "$trace(\pi_2) \in \mathcal{L}(\theta)$" *answers*
   | | | "*No*" **then**
15 | | | | **REJECT**
16 | | | **end**
17 | | | **else**
18 | | | | **ACCEPT**
19 | | | **end**
20 | | **end**
21 | **end**
22 **end**

Algorithm 1. $MissingAttack(\theta, \mathcal{S})$.

Correctness of Algorithm 1, i.e., the fact that it returns **ACCEPT** if, and only if, $\Delta_\theta^\mathcal{S} \neq \emptyset$, can be established as follows. Assume Algorithm 1 can return **ACCEPT**. Then, there is a way to choose $\pi \notin [\![\theta]\!]^\mathcal{S}$ (line 6), $\pi_1 \in [\![\theta]\!]^\mathcal{S}$ (line 10) and $\pi_2 \in [\![\theta]\!]^\mathcal{S}$ (line 14), such that $\pi.first = \pi_1.first$ and $\pi.last = \pi_2.last$ (line 2). By Proposition 1 and Definition 9, π is a missing attack, so that $\Delta_\theta^\mathcal{S} \neq \emptyset$. Reciprocally, if $\Delta_\theta^\mathcal{S} \neq \emptyset$, pick a missing attack $\pi \in \Delta_\theta^\mathcal{S}$. By Definition 9, $\pi \notin [\![\theta]\!]^\mathcal{S}$ and there must exist $\pi_1, \pi_2 \in [\![\theta]\!]^\mathcal{S}$ with $\pi.first = \pi_1.first$ and $\pi.last = \pi_2.last$, so Algorithm 1 can return **ACCEPT** by non-deterministically choosing these three paths, which concludes.

Corollary 1. *The MAE problem restricted to attack trees with operators ranging over* $\{$OR, SAND, AND$\}$ *is in* Σ_2^P *of the polynomial hierarchy.*

We recall that Σ_2^P is the class of problems that can be solved by a non-deterministic polynomial-time algorithm with queries to NP oracles [24].

We are able to establish that MAE is not easy[1] by showing its CO-NP-hardness.

Theorem 1. *The MAE problem is* CO-NP-*hard, even if we restrict to trees with operators ranging over* $\{$OR, SAND, AND$\}$ *only.*

Proof. We describe a reduction from TATM to MAE such that an instance of TATM is negative if, and only if, its reduction is a positive instance of MAE. Since Proposition 3 entails the NP-hardness of TATM, we easily conclude.

Let θ (over ACT) and $t = a_1 a_2 \dots a_n \in$ ACT* be an instance of TATM. Pick a fresh action symbol $\# \notin$ ACT. We define the attack tree $\theta^\# \overset{def}{=}$ OR$(\theta, \#)$ and the transition system $\mathcal{S}$ composed of $n + 1$ states $s_0, s_1, \dots s_n$ and containing only two paths: the path $\pi_t \overset{def}{=} s_0 \xrightarrow{a_1} s_1 \xrightarrow{a_2} \dots \xrightarrow{a_n} s_n$ and the path $s_0 \xrightarrow{\#} s_n$, as illustrated in Fig. 9. It is easy to prove that $\pi_t \in \Delta_{\theta^\#}^{\mathcal{S}}$ if, and only if, $t \notin \mathcal{L}(\theta)$, which concludes. $\qquad\square$

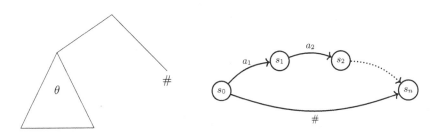

Fig. 9. Positive instance $(\theta^\#, \mathcal{S})$ of MAE obtained by reducing a negative instance $(\theta, a_1 \dots a_n)$ of TATM.

We do not know if MAE is Σ_2^P-hard, we unsuccessfully attempted to reduce the typical Σ_2^P-complete problem QBF$_2$ that asks if a quantified Boolean formula of the form $\exists x_1 \dots \exists x_n \forall y_1 \dots \forall y_m \varphi$, where φ is a formula over variables $x_1, \dots, y_m$, evaluates to true.

We now come back to proving Proposition 2 regarding the NP-completeness of TATM, and show it in the two next sections.

---

[1] This holds under the assumption that $P \neq NP$.

5.3   The NP-hardness of TATM

This section is dedicated to show that the TATM problem is NP-hard.

Proposition 3. *TATM is* NP-*hard, even if we discard weak operators* wSAND *and* wAND.

We consider the decision problem of *Packed Interval Covering* (PIC), which is NP-complete according to [22]. Let N be an integer. The PIC problem consists in deciding whether we can cover (in the classical sense) interval $[1, N]$ by selecting exactly one subinterval per pack of subintervals given as input. For instance, if the packs are $P_1 = \{[1, 6], [5, 9]\}$, $P_2 = \{[1, 3], [4, 6], [7, 7]\}$, $P_3 = \{[4, 4]\}$, we can cover $[1, 9]$ by selecting $[5, 9]$, $[1, 3]$ and $[4, 4]$, as illustrated in Fig. 10.

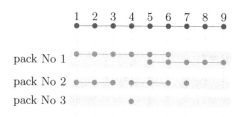

Fig. 10. An instance of the PIC problem.

Formally, we state the following definition.

Definition 12 (Packed Interval Covering (PIC) problem)

– *Input: an integer $N > 0$ and a family of finite sets $P_1, \ldots, P_M$ (packs) of subintervals of $[1, N]$.*
– *Output: are there subintervals $I_1 \in P_1, \ldots, I_M \in P_M$, such that $\bigcup_{k=1}^{M} I_k = [1, N]$?*

Proposition 4. *PIC is* NP-*complete.*

Arguing that PIC belongs to NP is easy. The certificate is simply a list of M sub-intervals of $[1, N]$ (given by their bounds), thus it is polynomial in the size of the packs input. It remains to verify for $i \in \{1, \ldots, M\}$ that interval of rank i belongs to pack P_i, and that the union of the chosen intervals covers $[1, N]$, which can all be performed in polynomial time. Regarding the NP-hardness of PIC, there is a polynomial reduction from the NP-complete problem (3, B2)-SAT [6], which is the restriction of 3-SAT where each variable has exactly two positive occurrences and two negative occurrences. The full proof of this reduction can be found in [22].

Back to the proof of Proposition 3, we exhibit a polynomial reduction of PIC into TATM, so TATM is NP-hard. This reduction relies on a transformation that,

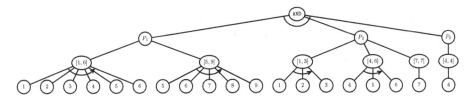

Fig. 11. The tree θ_0.

given an instance I of PIC, returns an instance I' of TATM, of size polynomial in the size of I, and such that I is a positive instance of PIC iff I' is a positive instance of TATM.

Instead of providing the full proof, we illustrate the idea on the example of the 3 packs $P_1 = \{[1,6],[5,9]\}$, $P_2 = \{[1,3],[4,6],[7,7]\}$, $P_3 = \{[4,4]\}$ to cover the whole interval $[1,9]$. Its corresponding instance of TATM is composed of the attack tree θ_0, depicted in Fig. 11 (we allow unary OR and SAND operators, as for nodes labeled P_3 and $[4,4]$ on the right-hand side of the tree, for instance), and the trace $t^0 = 123456789$.

Recall that this pack instance has solution $[5,9],[1,3],[4,4]$.

Consider the universal transition system $\mathcal{U}_{\{1,\dots,9\}}$, and the unique path π_0 whose trace is t^0, namely the path $s \xrightarrow{1} s \xrightarrow{2} s \dots \xrightarrow{9} s$.

According to the path semantics,

- π_0 is a parallel composition of paths $\pi_1 = s \xrightarrow{5} s \xrightarrow{6} s \xrightarrow{7} s \xrightarrow{8} s \xrightarrow{9} s$, $\pi_2 = s \xrightarrow{1} s \xrightarrow{2} s \xrightarrow{3} s$, and $\pi_3 = s \xrightarrow{4} s$, and
- π_1 (resp. π_2, π_3) belong to the path semantics of the left (resp. middle, right) subtree of the root node of θ_0.

As a consequence, $t^0 \in \mathcal{L}(\theta_0)$.

Notice that the proposed reduction yields attack trees that do not use weak operators, so that the NP-hardness of TATM holds even if we discard the use of weak operators in the input tree of the problem.

In order to achieve the proof of Proposition 2, namely to show that TATM is in NP, we have designed a non-deterministic polynomial-time algorithm that is explained in the next section.

5.4   The NP-membership of TATM

Proposition 5. *TATM is in* NP.

The proof of Proposition 5 is a direct consequence of the non-deterministic algorithm Algorithm 2 that reads the input trace while marking the nodes of the tree that have been "satisfied" by the prefix trace read so far. Algorithm 2 relies on the subroutine Algorithm 3, with initial call $check(\theta, t, \emptyset, startNodes(\theta), \emptyset)$.

Input: An attack tree θ (over ACT) and a non-empty trace $t \in$ ACT*
Output: ACCEPT if $t \in \mathcal{L}(\theta)$, **REJECT** otherwise.
1 $check(\theta, t, \emptyset, startNodes(\theta), \emptyset)$

Algorithm 2. $checkMembership(\theta, t)$.

Before explaining the subroutine $check(\theta, \gamma, \texttt{Must}, \texttt{May}, \texttt{Marked})$ (Algorithm 3), let us first fix the notation used.

We use $Nodes(\theta)$ to denote the set of nodes of the tree θ, and we write $root(\theta)$ for its root, and $Leaves(\theta)$ for its leaf node set. For a node $\gamma \in Nodes(\theta)$, we use self-explanatory notation: $children(\gamma)$, $parent(\gamma)$, and $ancestors(\gamma)$ (including γ itself).

We also consider $sib(\gamma)$ that denote the siblings of γ (excluding γ itself) and $rsib(\gamma)$ that refers to the right sibling of γ (if any). For $\gamma \in Leaves(\theta)$, we write $actionAt(\gamma)$ for the action that labels this node. Also, we may lift relevant notions to a set of nodes Γ, such as $actionAt(\Gamma)$, $ancestors(\Gamma)$.

We let $descendentleaves(\gamma)$ denote the set of leaf nodes of the subtree at node γ, and $startNodes(\theta)$ be the subset of nodes of θ whose labels are actions that the attacks may start with. For example, regarding the attack tree of Fig. 1, the set $startNodes(\theta)$ is composed of the leaf nodes that carry label either Bribe employee, or Blackmail employee, or Unlock with stolen key, or Force lock.

More formally, $startNodes(\theta)$ is defined by induction over θ and can be easily computed by a terminal recursive algorithm in linear time: $startNodes(a) = \{a\}$; for every OP $\in$ {OR, AND, wAND}, $startNodes(\texttt{OP}(\theta_1, \ldots, \theta_n)) = \bigcup(startNodes(\theta_i)$; for every OP $\in$ {SAND, wSAND}, $startNodes(\texttt{OP}(\theta_1, \ldots, \theta_n)) = startNodes(\theta_1)$.

We now explain Algorithm 3 that takes the following inputs:

- an attack tree θ (over ACT), according to Definition 1,
- a trace $t \in$ ACT*,
- a set of nodes Must that has to "progress" at next step, initialized to $\emptyset$ for the first call,
- a set of leaves May that may progress at next step, initialized as $startNodes(\theta)$,
- a set of nodes Marked that have already been "consumed" while reading trace t, initialized to $\emptyset$ for the first call.

The principle of Algorithm 3 is as follows:

1. Choose a set of leaves Γ inside May that contains Must, and the label of the chosen leaves matches the first action of the trace t.
2. Put those leaves γ in Marked.
3. Update Must, May and Marked accordingly by calling $propagate(\theta, \gamma, \texttt{Must}, \texttt{May}, \texttt{Marked})$ (Algorithm 4), for each γ in Γ.
4. Goto 1 with the updated sets Must, May, and Marked and the next action of the trace.

Input: A root node θ, a trace t, a set of nodes $\texttt{Must} \subseteq Nodes(\theta)$, a set of leaves
$\qquad$ $\texttt{May} \subseteq Leaves(\theta)$, a set of nodes $\texttt{Marked} \subseteq Nodes(\theta)$
Output: **ACCEPT** if $t \in \mathcal{L}(\theta)$, **REJECT** otherwise

```
1  if root(θ) ∈ Marked and t = ε then
2  |   ACCEPT
3  end
4  else
5  |   if May = ∅ or t = ε then
6  |   |   REJECT
7  |   end
8  |   else
9  |   |   CHOOSE ∅ ⊊ Γ ⊆ May ;
10 |   |   if actionAt(Γ) ⊄ {t(1), ⋆} then
11 |   |   |   REJECT
12 |   |   end
13 |   |   else
14 |   |   |   if Must ⊄ ancestors(Γ) then
15 |   |   |   |   REJECT
16 |   |   |   end
17 |   |   |   else
18 |   |   |   |   Must ← ∅; May ← May \ Γ; Marked ← Marked ∪ Γ ;
19 |   |   |   |   forall γ ∈ Γ with actionAt(γ) ≠ ⋆ do
20 |   |   |   |   |   propagate(θ, γ, Must, May, Marked)
21 |   |   |   |   end
22 |   |   |   end
23 |   |   |   return check(θ, t^{≥1}, Must, May, Marked)
24 |   |   end
25 |   end
26 end
```

Algorithm 3. $check(\theta, t, \texttt{Must}, \texttt{May}, \texttt{Marked})$.

The call to $propagate(\theta, \gamma, \texttt{Must}, \texttt{May}, \texttt{Marked})$ (Algorithm 4) allows us to update the sets $\texttt{May}$, $\texttt{Must}$ and $\texttt{Marked}$, with the consequences of marking the chosen leaves.

We set internal nodes as marked according to the path semantics: when a child of an OR node is marked so is this node, and when all the children of either of SAND, wSAND, AND, or a wAND node are marked, so is this node. Also, we put a leaf in $\texttt{May}$ when it gets enabled: namely, when the first child of a wSAND or SAND node is marked, its right sibling (if any) gets enabled.

Finally, we add nodes to $\texttt{Must}$ when they are expected to progress at the next step: for SAND nodes once a child is marked, the next child has to progress in the next step, and for an AND node one of its non-marked children has to progress in the next step. The propagation of all these constraints is recursive: at each newly marked node, $propagate$ is called again on this node.

The tricky part in Algorithm 4 is due to weak operators: we have to take into account that some actions in t may not be due to any leaf of the tree. This is

---

Input: A root node θ, Must $\subseteq Nodes(\theta)$, May $\subseteq Leaves(\theta)$,
   Marked $\subseteq Nodes(\theta)$, and node γ newly added to Marked
Output: Update of Must, May, and Marked

```
 1  if γ ≠ root(θ) then
 2  │   μ ← parent(γ);
 3  │   switch μ.OP do
 4  │   │   case OR do
 5  │   │   │   May ← May \ descendentleaves(μ);
    │   │   │       propagate(θ, Must, May, Marked, μ)
 6  │   │   end
 7  │   │   case SAND do
 8  │   │   │   if rsib(γ) exists then
 9  │   │   │   │   γ' ← rsib(γ); Must ← Must ∪ {γ'}; May ← May ∪ startNodes(γ')
10  │   │   │   end
11  │   │   │   else
12  │   │   │   │   propagate(θ, Must, May, Marked, μ)
13  │   │   │   end
14  │   │   end
15  │   │   case AND do
16  │   │   │   if sib(γ) ⊆ Marked then
17  │   │   │   │   propagate(θ, Must, May, Marked, μ)
18  │   │   │   end
19  │   │   │   else
20  │   │   │   │   Must ← Must ∪ {μ}
21  │   │   │   end
22  │   │   end
23  │   │   case wSAND do
24  │   │   │   if rsib(γ) exists then
25  │   │   │   │   μ ← μ.addchildwithlabel(⋆);
26  │   │   │   │   May ← May ∪ {μ}; May ← May ∪ {startNodes(rsib(γ))}
27  │   │   │   end
28  │   │   │   else
29  │   │   │   │   forall γ' ∈ children(μ) with actionAt(γ') = ⋆ do
30  │   │   │   │   │   May ← May \ {γ'}; μ.removechild(γ')
31  │   │   │   │   end
32  │   │   │   │   Marked ← Marked ∪ {μ}; propagate(θ, Must, May, Marked, μ)
33  │   │   │   end
34  │   │   │   case wAND do
35  │   │   │   │   if sib(γ) ⊆ Marked then
36  │   │   │   │   │   forall γ' ∈ children(μ) with actionAt(γ') = ⋆ do
37  │   │   │   │   │   │   May ← May \ {γ'}; μ.removechild(γ')
38  │   │   │   │   │   end
39  │   │   │   │   │   Marked ← Marked ∪ {μ}; propagate(θ, Must, May, Marked, μ)
40  │   │   │   │   end
41  │   │   │   │   else
42  │   │   │   │   │   μ ← μ.addchildwithlabel(⋆); May ← May ∪ {μ}
43  │   │   │   │   end
44  │   │   │   end
45  │   │   end
46  │   end
47  end
```

Algorithm 4. $propagate(\theta, \text{Must}, \text{May}, \text{Marked}, \gamma)$.

done by dynamically adding and removing artificial leaves to wSAND and wAND nodes, with special label ⋆, so that these new leaves can be chosen to validate the Must requirements. These added leaves somehow correspond to an "ignore" instruction while reading the trace: if there are some weak operators in the tree and we encounter an action in t that is not a leaf of θ, the algorithm can simply choose to ignore it and keep on with the next action of t, while validating the Must requirement on the parent node.

Finally, trace t is in $\mathcal{L}(\theta)$ if we manage to read the whole trace, and if the root of the tree is marked in the end. Otherwise, the trace is rejected.

5.5   TATM Implementation

As illustrated by Algorithm 1, answering the MAE problem involves solving the TATM problem. To support our approach, we have thus implemented a proof of concept tool automating the solving of TATM: it checks the membership of each trace in the language of the specified tree. The implementation is open source and is available online at https://github.com/yanntm/Abat, visualized in Fig. 13.

The prototype includes a small Xtext based editor that allows to specify an attack tree and a set of traces. Its current version supports only trees without

```
tree = AND (
OR ( SAND ("a1","a2","a3","a4","a5","a6") , SAND("a5","a6","a7","a8","a9")),
OR ( SAND ("a1","a2","a3") , SAND("a4","a5","a6"), "a7" ),
"a4"
);
```

The inputs traces are successively:

```
trace interval_1_1 = "a1";
trace interval_1_2 = "a1","a2";
trace interval_1_3 = "a1","a2","a3";
trace interval_1_4 = "a1","a2","a3","a4";
trace interval_1_5 = "a1","a2","a3","a4","a5";
trace interval_1_6 = "a1","a2","a3","a4","a5","a6";
trace interval_1_7 = "a1","a2","a3","a4","a5","a6","a7";
trace interval_1_8 = "a1","a2","a3","a4","a5","a6","a7","a8";
trace interval_1_9 = "a1","a2","a3","a4","a5","a6","a7","a8","a9";
trace interval_1_10 = "a1","a2","a3","a4","a5","a6","a7","a8","a9","a10";
```

For each trace, the tool respectively answers:

```
Trace "interval_1_1" rejected
Trace "interval_1_2" rejected
Trace "interval_1_3" rejected
Trace "interval_1_4" rejected
Trace "interval_1_5" rejected
Trace "interval_1_6" accepted
Trace "interval_1_7" accepted
Trace "interval_1_8" rejected
Trace "interval_1_9" accepted
Trace "interval_1_10" rejected
```

Fig. 12. Run of the tool https://github.com/yanntm/Abat.

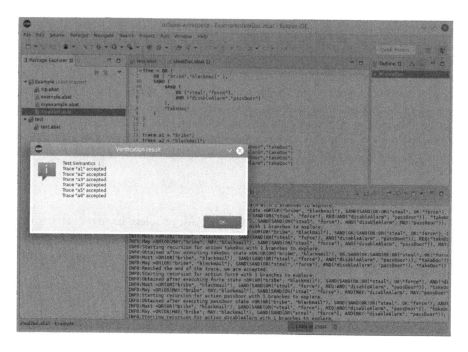

Fig. 13. Interface of our prototype https://github.com/yanntm/Abat.

weak operators, but we will soon be able to consider arbitrary trees by extending the backtracking approach to the whole Algorithm 3.

We illustrate how our implementation for solving TATM works on the example of the tree illustrated in Fig. 11.

The input tree and traces, as well as the answer from the tool, are shown in Fig. 12.

6   Conclusion and Future Work

We have proposed a framework where attack trees are interpreted according to a model of the system, thus yielding their path semantics, but mostly displaying a natural notion of missing attacks. We then have considered the decision problem of the existence of a missing attack (MAE), which is highly pertinent for attack tree designers.

It should be noticed that our notion of missing attack relies on the model of the system and not on the system itself, just as the model checking principle in system verification makes the model design an upstream issue.

We insist on the robustness of the proposed approach for our models of systems: those are transition systems that allow for non-deterministic behavior. This way, the attacker executing an action may not control its effects, which captures the idea that the attacker interacts with some environment (seen as an abstract opponent that solves the non-determinism – a very standard way of modeling in formal methods). The attacks are thus sequences of actions that the attacker can entirely perform, if the environment does not prevent him from doing so.

Also, regarding our formal setting, we have equipped attack trees not only with standard operators, but also with weak variants of those, allowing more flexibility in the specification and getting much closer to our intuition when reading informal trees developed in practice.

From our path semantics, and by considering the universal system where any sequence of actions can be executed, we have also defined a trace semantics for attack trees, thus offering an interpretation of attack trees on their own, and exhibiting a formal notion to investigate the missing attack problem MAE. In particular, we studied the trace attack tree membership problem TATM and showed it is NP-complete. Noticeably, the NP-hardness proof for TATM, even for trees with no weak operators, resorts to a newly defined combinatorial problem PIC that, although very natural, has never been considered in the literature.

Next, relying on an NP oracle that answers TATM, we could design non-deterministic polynomial-time algorithm to solve MAE when weak operators are discarded. The algorithm guesses three paths and makes three independent calls to this oracle for each path, which shows its membership in the complexity class Σ_2^P of the polynomial hierarchy [24] (containing the classes NP and co-NP).

On our way to develop tools for attack tree users, we have implemented the NP oracles: on the basis of the non-deterministic algorithms described in Sect. 5, we have coded their deterministic version as a backtracking algorithm. The tool is freely available and open source. It is for now mostly dedicated to educational purposes for better understanding the chosen semantics of the operators.

There are several directions to pursue this work. First, complexity bounds need to be made tighter. While we know that MAE is in Σ_2^P and that it is as hard as any CO-NP problem, we still need to fill this gap. Next, we should address the complexity of (full[2]) MAE, which requires to provide a bound on the three paths guessed by Algorithm 1 in the general case, rather than for trees with no weak operators.

Obviously the aforementioned missing complexity results would shed light on the difficulty of synthesizing missing attacks, or provide hints for subclasses of instances where the MAE problem might become simpler.

Moreover, it could be interesting to investigate cases where the TATM problem is easier. For example, when the tree does not contain any weak operators and actions appear at most once in the tree, the complexity of TATM becomes linear since there is only one way of interpreting an action in the tree.

Finally, we wish to study variants of the path/trace semantics of attack trees for weak operators. After all, weak operators offer a way to abstract from intermediate sequences of actions. These actions are neglected by the designer for some reason that need to be better understood. We may refine the current semantics by considering that these actions should be other than those occurring

---

[2] In the full MAE problem, all (strong and weak) operators are allowed.

in the tree. The definition of the path semantics can be adapted accordingly, and does not change our current results, while giving a hope to obtain complexity upper bounds for MAE with arbitrary trees.

References

1. Amenaza: SecurITree (2001–2013). http://www.amenaza.com/
2. Audinot, M., Pinchinat, S., Kordy, B.: Is my attack tree correct? In: Foley, S.N., Gollmann, D., Snekkenes, E. (eds.) ESORICS 2017. LNCS, vol. 10492, pp. 83–102. Springer, Cham (2017). https://doi.org/10.1007/978-3-319-66402-6_7
3. Audinot, M., Pinchinat, S., Kordy, B.: Guided design of attack trees: a system-based approach. In: CSF, pp. 61–75. IEEE Computer Society (2018)
4. Audinot, M., Pinchinat, S., Schwarzentruber, F., Wacheux, F.: Deciding the non-emptiness of attack trees. In: Cybenko, G., Pym, D., Fila, B. (eds.) GraMSec 2018. LNCS, vol. 11086, pp. 13–30. Springer, Cham (2019). https://doi.org/10.1007/978-3-030-15465-3_2
5. Baier, C., Katoen, J.: Principles of Model Checking. MIT Press, Cambridge (2008)
6. Berman, P., Karpinski, M., Scott, A.D.: Approximation hardness of short symmetric instances of MAX-3SAT. Electronic Colloquium on Computational Complexity (ECCC) 10(049) (2003). http://eccc.hpi-web.de/eccc-reports/2003/TR03-049/index.html
7. EAC Advisory Board and Standards Board: Election Operations Assessment - Threat Trees and Matrices and Threat Instance Risk Analyzer (TIRA) (2009). https://www.eac.gov/assets/1/28/Election_Operations_Assessment_Threat_Trees_and_Matrices_and_Threat_Instance_Risk_Analyzer_(TIRA).pdf
8. Gadyatskaya, O., Harpes, C., Mauw, S., Muller, C., Muller, S.: Bridging two worlds: reconciling practical risk assessment methodologies with theory of attack trees. In: Kordy, B., Ekstedt, M., Kim, D.S. (eds.) GraMSec 2016. LNCS, vol. 9987, pp. 80–93. Springer, Cham (2016). https://doi.org/10.1007/978-3-319-46263-9_5
9. Gadyatskaya, O., Jhawar, R., Mauw, S., Trujillo-Rasua, R., Willemse, T.A.C.: Refinement-aware generation of attack trees. In: Livraga, G., Mitchell, C. (eds.) STM 2017. LNCS, vol. 10547, pp. 164–179. Springer, Cham (2017). https://doi.org/10.1007/978-3-319-68063-7_11
10. Hong, J.B., Kim, D.S., Chung, C., Huang, D.: A survey on the usability and practical applications of Graphical Security Models. Comput. Sci. Rev. **26**, 1–16 (2017)
11. Isograph: AttackTree+ (2004–2005). http://www.isograph-software.com/atpover.htm
12. Ivanova, M.G., Probst, C.W., Hansen, R.R., Kammüller, F.: Attack tree generation by policy invalidation. In: Akram, R.N., Jajodia, S. (eds.) WISTP 2015. LNCS, vol. 9311, pp. 249–259. Springer, Cham (2015). https://doi.org/10.1007/978-3-319-24018-3_16
13. Jhawar, R., Kordy, B., Mauw, S., Radomirović, S., Trujillo-Rasua, R.: Attack trees with sequential conjunction. In: Federrath, H., Gollmann, D. (eds.) SEC 2015. IAICT, vol. 455, pp. 339–353. Springer, Cham (2015). https://doi.org/10.1007/978-3-319-18467-8_23
14. Jürgenson, A., Willemson, J.: Computing exact outcomes of multi-parameter attack trees. In: Meersman, R., Tari, Z. (eds.) OTM 2008. LNCS, vol. 5332, pp. 1036–1051. Springer, Heidelberg (2008). https://doi.org/10.1007/978-3-540-88873-4_8

15. Kordy, B., Piètre-Cambacédès, L., Schweitzer, P.: DAG-based attack and defense modeling: don't miss the forest for the attack trees. Comput. Sci. Rev. **13–14**, 1–38 (2014)

16. Kordy, B., Wideł, W.: On quantitative analysis of attack–defense trees with repeated labels. In: Bauer, L., Küsters, R. (eds.) POST 2018. LNCS, vol. 10804, pp. 325–346. Springer, Cham (2018). https://doi.org/10.1007/978-3-319-89722-6_14

17. Mantel, H., Probst, C.W.: On the meaning and purpose of attack trees. In: CSF, pp. 184–199. IEEE Computer Society (2019)

18. Mauw, S., Oostdijk, M.: Foundations of attack trees. In: Won, D.H., Kim, S. (eds.) ICISC 2005. LNCS, vol. 3935, pp. 186–198. Springer, Heidelberg (2006). https://doi.org/10.1007/11734727_17

19. National Electric Sector Cybersecurity Organization Resource (NESCOR): Analysis of selected electric sector high risk failure scenarios, version 2.0 (2015). http://smartgrid.epri.com/doc/NESCOR

20. Pinchinat, S., Acher, M., Vojtisek, D.: Towards synthesis of attack trees for supporting computer-aided risk analysis. In: Canal, C., Idani, A. (eds.) SEFM 2014. LNCS, vol. 8938, pp. 363–375. Springer, Cham (2015). https://doi.org/10.1007/978-3-319-15201-1_24

21. Pinchinat, S., Acher, M., Vojtisek, D.: ATSyRa: an integrated environment for synthesizing attack trees. In: Mauw, S., Kordy, B., Jajodia, S. (eds.) GraMSec 2015. LNCS, vol. 9390, pp. 97–101. Springer, Cham (2016). https://doi.org/10.1007/978-3-319-29968-6_7

22. Saffidine, A., Cong, S.L., Pinchinat, S., Schwarzentruber, F.: The Packed Interval Covering Problem is NP-complete. CoRR abs/1906.03676 (2019). http://arxiv.org/abs/1906.03676

23. Schneier, B.: Attack trees. Dr. Dobb's J. **24**(12), 21–29 (1999)

24. Stockmeyer, L.J.: The polynomial-time hierarchy. Theoret. Comput. Sci. **3**(1), 1–22 (1976)

25. Vigo, R., Nielson, F., Nielson, H.R.: Automated generation of attack trees. In: CSF, pp. 337–350. IEEE Computer Society (2014)

Optimizing System Architecture Cost and Security Countermeasures

Sahar Berro[ID], Ludovic Apvrille[ID], and Guillaume Duc[✉][ID]

LTCI, Télécom Paris, Institut Polytechnique de Paris, Palaiseau, France
{sahar.berro,ludovic.apvrille,guillaume.duc}@telecom-paris.fr

Abstract. The design of an embedded system is built on a trade-off between its performance and its cost. Nowadays, the security threats that target most of the embedded systems introduce a new factor in this trade-off: the security level of the system. So system architects must consider, during the design, the different attacks that target the system and the possible countermeasures, and their costs. In this article, we present a methodology to help designers explore different countermeasures and evaluate their impact on the cost of the architecture and the probability of success of an adversary. This methodology is based on extended and formalized Attack-Defense Trees that allow to assess the impact of countermeasures on system components and attacks. We use propagation rules to characterize a main attack from its different steps, and we formalize the trade-off between security and cost by an optimization problem between attack probability and total architecture cost.

Keywords: Attack-Defense Tree · Security of embedded system · Countermeasures

1 Introduction

System-level embedded system design—e.g. with model based approaches—is a common practice that unfortunately frequently ignores cyber-security aspects. Thus, usual system-level approaches rather target safety and performance aspects [11,24]. Security can be important (i) by itself because of e.g. privacy concerns and (ii) because it could impact safety and performance [1].

Model-based approaches usually rely on security requirements diagrams and on attack (defense) trees to capture security aspects [3,19,22]. Attack trees structure attacks in a way that intends to help designers select the relevant countermeasures that can prevent the root attacks of trees. Yet, countermeasures are strongly linked to system architectural aspects for two reasons. First, a countermeasure may be implemented only in a given architecture. For instance, a powerful crypto accelerator can surely not be implemented in a very low power device. A similar example can be found in the EVITA architecture where Hardware Security Modules are added to Electronic Control Units: since some of them are expected to be of very low cost, different versions of the HSM (light, medium,

© Springer Nature Switzerland AG 2019
M. Albanese et al. (Eds.): GraMSec 2019, LNCS 11720, pp. 50–67, 2019.
https://doi.org/10.1007/978-3-030-36537-0_4

full) have been defined. Second, a countermeasure depends on the already existing components of an architecture. For instance, if an attack consists in exploiting a known CVE for a given Operating System O1, replacing O1 by another operating system O2 depends on the fact that O2 exists for the processors of the target architecture. Finally, identifying the right countermeasures is an optimization process that should take into account both the interest in resisting to attacks and the cost it implies on the architecture.

SysML-Sec [2,4] has already been proposed to handle in the same framework security requirements, attack trees and architecture design (in particular platform cost). In particular, previous contributions have shown the relations between attacks and architectural elements. This paper enhances previous contributions with the definitions on how an optimal architecture could be found according to attacks likeliness and countermeasure cost. For this, the paper introduces a new approach based on Attack-Defense Trees (ADT) enhanced with formalized links to architectural elements.

The rest of the paper is organized as follows. Section 2 details the previous works related to our work. Section 3 presents the context of our work. Section 4 describes our main contributions. Section 5 explains some choices we made. Section 6 presents the application of our methodology to a case study. Section 7 discusses future directions for our work.

2   State of the Art

The basic formalism of Attack Trees (AT) was first introduced by Schneier in [22].

Mauw and Oostdjik in [20] proposed and alternative formalism for the one presented by Schneier. Their contribution consists in associating a set of *mincuts* to the root attack.

Contributions of Jürgenson et al. in [13] improved the semantics of Mauw and Oostdjik in [20] and introduced an exact and consistent set of computational rules to determine attackers' expected outcomes. In particular, their contribution takes into account parameters in leaves: attack cost, probability of success, expected penalty on the attacker if the attack was unsuccessful and the expected penalty on the system in case the attack was successful. They also used a global parameter "gains" to evaluate the benefit of the attacker in case it could achieve the root attack (and not only elementary attacks).

Audinot et al. in [7] showed the complexity results for three notions of the soundness of an attack tree: admissibility, consistency and completeness. They also show how the tree operators influence complexity results.

Other proposals suggest to make a combine use of attack trees and fault trees. Steiner and Liggesmeyer [23] have extended the qualitative and quantitative safety analysis to take in consideration the influence of security problems on the safety of a system. They introduced "SECFT" (that stands for Security Event Component Fault Tree), a component fault tree that contains both safety and security events. They conduct their security analysis using likelihood of occurrence for security events (attacks) and the probability of occurrence for faults.

However, none of these contributions take into account defense nor protection mechanisms. Bistarelly et al. [8] introduce the notion of defense tree DT. They proposed to model attackers and defenders using game theory in order to find the set of countermeasures that has the most effective cost. To evaluate the return on attack ROA, they used the expected gain of a successful attack, its cost and the augmented cost caused by the use of a countermeasure (revised cost). To evaluate the return on security investment ROI, they used the annual financial loss caused by a threat, the annual number of occurrence of a threat, the cost of the countermeasure and the impact of threats on these countermeasures. Kordy et al. in [16] gave a formalism where attack trees used in [9,14,21] and defense trees introduced by Bistarelly are covered in a single framework: attack-defense tree ADT. They developed in [15] ADTool that supports quantitative and qualitative analysis of ADTs and their instances.

Edge et al. [9] proposed to use protection trees along with attack trees to determine the protections needed for computer networks in homeland security. They also defined a set of rules to evaluate metrics associated with the leaves of both trees. For attack trees, they used (only) the probability of success and the cost of attacks in order to evaluate the impacts on the risk of a system.

Ji et al. [12] analyzed the performance of ADTs by assigning additional parameters to each attack and countermeasure nodes: cost, impact on the system and success probability for attacks, and cost for countermeasures. These parameters are then used for evaluating the revised impact on the system and the revised cost of the attack after the countermeasure is deployed. Then, they analyze the performance of the ADT by considering the ROI and the ROA.

Fraile et al. [10] presented a case study where they modeled and analyzed threats on ATMs using ADTs. They used occurrence probability for each attack node to derive the likelihood for the overall root ATM attacks. The probability of occurrence of a refined attack node was calculated by taking in consideration whether or not this node is counteracted by a defense mechanism.

Kordy and Widel [18] defined a set of rules to conduct a quantitative analysis on an ADT with repeated labels. As metrics, they used the minimal cost for executing an attack, the maximal damage caused by an attack, the minimal skill level required to execute an attack, the minimal number of experts needed to mount an attack and the satisfiability for the defender.

Finally, some contributions rely on attack trees to conduct only a security analysis and thus determine the impact of attacks on a given system. Some others rely on fault and attack trees in order to evaluate how attacks may cause faults and impact system safety. Some other proposals propose to rely on ADTs to represent how attacks are countered and thus they can analyze the interactions between the gain of a successful attack and the security investment cost. A state of the art on attack and defense modeling approaches for security is presented in [17] and a survey on graphical security models that gives an overview of their developements, complexity analysis and application has been provided in [5].

However none of these approaches have considered system architecture while evaluating attacks, countermeasures and their costs. The rest of the paper explains how we can efficiently iterate through attacks, countermeasures and system architectures in order to setup a "good" solution.

3   Context

Designing a system architecture is an incremental process. Each time a new component is added or modified (it could be a new hardware or software component, or simply a change of configuration of an existing component), new attacks may be possible on the system, and attacks that have been previously captured in attack trees may not be possible anymore. To counter all possible attacks, security engineers employ a set of countermeasures. These countermeasures could be implemented by modifying existing software, introducing new software components, adding hardware components, or changing the configuration of hardware components, taking us back to identify new attacks and so on. Obviously, the use of countermeasures is expected to decrease the impact of at least one attack, yet it could introduce new attacks and increase the cost of the platform.

The design of a system usually involves trade-offs between safety, security, performance and cost. The approach proposed in this paper aims at helping system designers to perform trade-offs between security level and platform cost, based on enhanced and formalized ADTs.

4   Contributions

4.1   Main Definitions: Adversary, Attack, Countermeasure

Since attacks are related to the architectural components, our contribution first consists in introducing a new formalism for ADTs, where not only attacks and countermeasures are modeled, but also the architecture components. In other terms, components where attacks may occur and countermeasures components are represented in the ADT in order to support our optimization based on attacks - countermeasures - architecture.

In order to conduct a security analysis using an ADT, first metrics are associated to the child nodes of the tree. Then, a set of propagation rules are defined. Finally, based on a bottom-up approach, metrics are evaluated from the leaf nodes to the root attack using these rules.

Attackers and Attacks

Definition 1. *A **Malicious attacker** M is a 2-uple (R_M, E_M) where:*

1. *R_M represents the set of resources of the adversary, i.e. hardware equipment, financial and time resources...*
2. *E_M represents the expertise of the adversary. E_M can be expressed as a value $E_M \in [0..1]$ (for instance) or as a label $E_M \in \{beginner, intermediate, expert\}$ (as long as the set of labels is ordered).*

The hardware equipment includes computation capabilities, electronic equipment such as soldering stations, boards, microscopes, lasers... Time represents the manpower and the time frame that the attacker has. For simplicity reasons, R could also be simply represented as a sum of all resources, thus simply representing the financial capabilities of an attacker.

Definition 2. *An **attack** A is a 3-uple (l, R_A, E_A) where:*

1. *l is a labeling function.*
2. *R_A represents the minimal set of resources that are necessary to perform this attack.*
3. *E_A represents the minimal expertise necessary for an adversary to perform this attack.*

Countermeasure. The objective of a countermeasure is to make an attack more difficult, i.e. to increase its cost or the necessary expertise.

Definition 3. *A **countermeasure** CM is a 4-uple $(l, C_{CM|SA}, I = \{(A, R, E)\}, N = \{A\})$ where:*

1. *l is a labeling function.*
2. *$C_{CM|SA}$ represents the cost of this countermeasure in a given architecture.*
3. *I is a set of 3-uple (A, R, E) that represents for a given attack A how it impacts its resources R and its expertise E. By definition, a countermeasure must increase either R or E of A. R and E may depend on the cost of the countermeasure.*
4. *N is a set of attacks that must be performed in addition to the existing attacks to circumvent the countermeasure. The parameters of these attacks may depend on the cost of the countermeasure.*

Countermeasures either impact existing attacks by making them more difficult to be realized (e.g. a masking countermeasure against a side-channel attack makes this attack more difficult by requiring a second-order attack) or introduce new attacks that must be performed in addition to existing ones to achieve the same result (e.g. a bus probing attack can be prevented by using encryption mechanisms; but this countermeasure can be circumvented by retrieving the key used to encrypt data transiting on the bus; at the end, the adversary must perform both attacks to retrieve the data in cleartext).

The cost of a given countermeasure may vary (for instance a countermeasure may have different levels of security but at different costs). Thus, the impact of a countermeasure may vary depending on the cost.

4.2   Success of Attacks

The success of an attack, with regards to an attacker, depends on the resources and expertise of an attacker. It can be defined as follows:

Definition 4. *The function **success** (attack, attacker) → true/false is a function that returns true if the attacker has the necessary resources and expertise to perform the attack, false otherwise: $success(A, M) = R_M \geq R_A \wedge E_M \geq E_A$.*

Considering a set (or population) of malicious attackers $\mathcal{M}$, it is now possible to define the probability of success of an attack.

Definition 5. *The **probability of success of an attack** A by a set of malicious attackers $\mathcal{M}$ is defined as:*
$$\mathcal{P}_{A/\mathcal{M}} = \frac{|\{M \in \mathcal{M} | success(A,M) = true\}|}{|\mathcal{M}|}$$

We can note that the success of a given attack with a given attacker and so the probability of success of a given attack depend on the countermeasures implemented on the system because they impact the resources or the level of expertise needed to carry off the attack or add additional attack steps that must be performed.

4.3 Attack Tree

An Attack Tree (AT) is a conceptual multi-level diagram used to describe the security of a system. It contains a root attack, intermediate and leaf attacks as well as some nodes of operators (usually AND and OR). This multi-layered approach allows to represent the different possible attack scenarios to reach the root one.

An Attack-Defense Tree (ADT) adds defense mechanisms to the set of attacks.

In this paper, we consider the following definition for our ADT:

Definition 6. *An Attack-Defense Tree for a given system architecture SA **ADT-SA** is a 5-uple*
$(OP, A_{root}, ATK, D, COMP)$ where:

1. *OP is the set of the different operators.*
2. *A_{root} is the root attack i.e. the main goal of the adversary.*
3. *ATK is the set of attack i.e. intermediate and leaf attacks.*
4. *D is the set of defense mechanisms i.e. the available countermeasures used to counter attacks.*
5. *COMP is the set of all the hardware and software components of the system architecture.*

Note that ATK, D and $COMP$ are not independent since a countermeasure may be implemented only if specific components are part of the system architecture. Similarly, a component or a set of components may introduce new attacks.

4.4 Example

To illustrate the concept of ADT-SA, we consider a toy embedded system with a processor, an external RAM (not embedded inside the processor), a bus

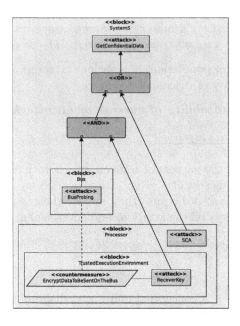

Fig. 1. Example of an ADT-SA

between the processor and the RAM. An operating system and an application that performs cryptographic operations (such as encryption) using a secret key run on this system.

Figure 1 shows one ADT-SA of the considered system. For simplicity reasons, we did not represent the whole system architecture. By the way, an ADT-SA does not require to model the whole architecture of the system, but only the part related to the ADT-SA under study.

In the considered ADT-SA, the main objective of the attacker (attack *get-ConfidentialData*) is to retrieve confidential data (in our case, the secret key manipulated by the application). To perform this attack, the attacker can either (due to the OR operator) perform a bus probing attack (attack *ProbingTheBus* that targets the component *Bus*) or a side-channel attack against the *Processor* during the manipulation of the secret key.

A countermeasure, *EncryptDataOnTheBus* (which consists in using a Trusted Execution Environment to encrypt all the data that are sent outside of the processor, for instance to the memory), can be used to prevent the bus probing attack. With this countermeasure, the adversary can probe the bus but it will only retrieve encrypted data that are useless unless it recovers the key used to encrypt the bus. So, this countermeasure adds an additional attack step (*RecoverKey*) that must be performed by the adversary, in addition to retrieving the encrypted data by bus probing, to achieve its goal.

4.5 Analysis

Describing attack scenarios in a graphical way is not the only objective of attack trees. They can indeed be used to perform a quantitative analysis with respect to given metrics called *attributes*. Attributes are metrics assigned to the basic actions in the tree (leaf nodes) and are used in a bottom-up evaluation. The bottom-up procedure consists in propagating attributes from leaf nodes to the root of the tree by applying appropriate operations to the intermediate operators connecting different intermediate attack steps. In this paper we will call these operations *propagation rules*.

Metrics and Propagation Rules. Our approach considers the following metrics:

- Attacks: minimal resources R and E necessary to perform the attack.
- Countermeasure: cost $C_{CM|SA}$ to implement the countermeasure in the corresponding system architecture, the increase of the minimal resources ΔR and of the minimal expertise required ΔE induced by the addition of this countermeasure in the corresponding system architecture (these increases can be directly caused by the countermeasure or indirectly by adding some attack steps needed to circumvent the countermeasure).
- System architecture: cost C_{arch}.

We now show how these metrics can be propagated from leaf nodes to higher nodes in the ADT-SA when using conjunction (denoted by AND) and disjunction (denoted by OR) operators.

1. **Minimal resources and expertise**
   Let us suppose that we have a root attack A_{root} (*AttackRoot*) with the following characteristics $(l_{root}, R_{root}, E_{root})$ and p other refined attacks such as $\forall i \in \{1, ..., p\}$, A_i (*Attacki*) is characterized by (l_i, R_i, E_i). Let us suppose as well that a malicious attacker M with (R_M, E_M) wants to perform A_{root} in the system.
   Figure 2 represents the AND operator where p attack steps are required to be performed in order to achieve the root attack. Figure 3 illustrates the OR where at least one attack step among p elementary attacks need to be achieved in order to realize the attack root.
   - **AND operator** The attacker M has to perform all the p attacks in order to realize A_{root}. Thus, its resources R_M must be at least equal to the sum of the resources needed to perform each attacks (we suppose here that all the attacks are totally independent, this supposition is discussed later). In other words, to succeed and realize its goal, its resources must be greater or equal to $\sum_{i=1}^{p} R_i$. Regarding the expertise, to mount A_{root}, the level of expertise of the attacker must be greater or equal to the maximum level of expertise required by the different attack steps: $\max_{i=1}^{p} E_i$.
     Therefore, $R_{root} = \sum_{i=1}^{p} R_i$ and $E_{root} = \max_{i=1}^{p} E_i$.

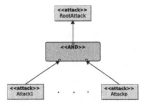 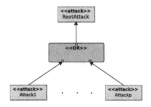

Fig. 2. AND operator   **Fig. 3.** OR operator

– **OR operator** To achieve A_{root}, the attacker M, characterized by (R_M, E_M), has to perform at least one of the p attacks.

If $\forall i \in \{1, ..., p\}$, $R_i > R_M$ or $E_i > E_M$, the attacker M will not be able to achieve A_{root}. Otherwise, among the attacks it has a sufficient level of expertise to perform ($Q = \{i | E_i \le E_M\}$), it will choose the one which requires the minimum resources (A_k such as $\forall i \in Q, R_k \le R_i$). In this case, $R_{root} = R_k$ and $E_{root} = E_k$.

We note that in this situation, the propagation rule depends on the attacker.

2. **Countermeasures and the increase of the minimal resources and level of expertise**

A countermeasure makes an attack more difficult to be mounted due to the fact that it introduces either new attack steps (which indirectly increase the resources and expertise needed to perform the attack thanks to the propagation rules defined previously) or directly increase the resources and expertise of the original attack.

Let us consider Fig. 4. For the sake of simplicity, we will use an arbitrary attack-defense tree with 2 leaf attacks connected with an OR to illustrate the evaluation of the impact of countermeasures. Note that the OR operator that connects A_1 and A_2 could be replaced by an AND operator, provided that we then use the corresponding propagation rules.

In this illustration, to achieve A_{root}, an attacker needs to perform either A_1 or A_2. The red segment between a countermeasure CM and an attack A, means that A is introduced by CM. The arc between CM_3 **and** CM_4 means that they are both needed to counter A_2 (they could have been represented by only one countermeasure that includes both). A_1 is prevented by CM_1 **or** CM_2. We will study how these two combinations of countermeasures (OR and AND) impact the attack.

In this paper, we will suppose that all the added countermeasures to the system are independent.

– **OR operator** Let us suppose that the attacker chooses to perform A_1 to achieve A_{root} and that the countermeasure CM_1 (respectively CM_2) adds an additional attack step A_{CM_1} (respectively A_{CM_2}). Figure 5 shows the developed tree with these two new attack steps (we did not represent A_2 countermeasures on the graph to make the tree less complicated).

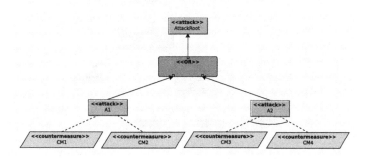

Fig. 4. Example of the impact of countermeasures

Since A_1 is prevented by CM_1 or by CM_2, to reach its goal, the attacker M has to bypass both countermeasures by performing the two new attacks A_{CM_1} and A_{CM_2}. Thus, to perform A_{root}, M must perform A_1, A_{CM_1} and A_{CM_2}, instead of just A_1.

Therefore, according to the propagation rules defined above, $R_{root} = R_1 + R_{CM_1} + R_{CM_2}$ and $E_{root} = \max(E_1, E_{CM_1}, E_{CM_2})$.

- **AND operator** Let us now suppose that the attacker M chooses to perform A_2 to achieve A_{root} and that the countermeasure CM_3 (respectively CM_4) adds an additional attack step A_{CM_3} (respectively A_{CM_4}). Figure 6 shows the developed tree with these two new attack steps (we did not represent A_1 countermeasures on the graph to make the tree less complicated).

  A_2 is prevented by CM_3 and CM_4 (both of them are required). Therefore, the attacker M has to bypass at least one of these countermeasures to reach its goal and achieve A_{root}. To do so, it must perform either A_{CM_3} or A_{CM_4}. As described before when the adversary can choose, among the attacks it has the level of expertise required, it will choose the one which requires the less amount of resource.

3. **System architecture.** System architecture is a description of the design of this system. It represents a plan of the interrelations between its existing or future components and subsystems. Initially this representation is general. As we go deeper into details, it can be refined to more concrete description. To prevent attacks, security experts use countermeasures. Hence, the cost of the whole system architecture C_{arch} is equal to the sum of the cost of all its components $C_{Total_{Comp}}$ and the cost of all its countermeasures $C_{Total_{CM}|SA}$:

$$C_{arch} = C_{Total_{Comp}} + C_{Total_{CM}|SA} \qquad (1)$$

However, using countermeasures may have some secondary effects: it may introduce new attacks and increase the cost of the platform. Moreover, system engineer usually fix a budget that should be respected while constructing

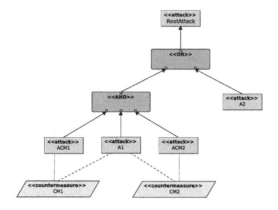

Fig. 5. Example of an OR operator applied on countermeasures

the system. Furthermore, designing a system require adjustments between performance, security/safety and cost. Thus, we need to find the set of countermeasures such that (i) Eq. 1 respects the pre-defined budget and (ii) leads to the minimal probability of success for a given attack. We will describe this problematic with the following optimization problem.

4.6   The Optimization Problem

Definition 7. *Let* ***S*** *be the set of the possible countermeasures used in the system and let* $\mathcal{P}(\mathbf{S})$ *be the powerset of* S.

Definition 8. *We define the function* C_{arch} : $\mathcal{P}(S) \rightarrow \mathbb{R}^+$ *as* $C_{arch}(x) = C_{Total_{Comp_x}} + C_{Total_x|SA}$, *where:*

- $C_{Total_{Comp_x}}$ *represents the cost of all the components of the system. That cost depends on the countermeasures that are implemented (x), since a countermeasure may need specific hardware components, e.g a powerful processor, an hardware cryptographic accelerator, etc.*
- $C_{Total_x|SA}$ *represents the cost of all the countermeasures used in system architecture SA.*

This function represents the total cost of the system depending on the implemented countermeasures (possibly none).

Definition 9. *We define the function* $P : \mathcal{P}(S) \rightarrow [0,1]$ *as* $P(x) = \mathcal{P}_{A_{root}|x}/\mathcal{M}$ *i.e. the probability of success of the root attack given a population of malicious attackers* $\mathcal{M}$ *and the countermeasures* x *implemented on the system.*

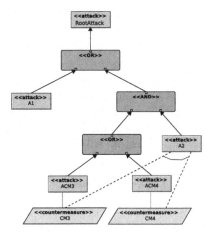

Fig. 6. Example of an AND operator applied on countermeasures

The problem is to find the set of countermeasures x that minimizes $P(x)$ with the constraint $C_{arch}(x) \leq B$, where B is the pre-defined budget fixed by the system engineers. This formalizes the trade-off that the system architect has to perform between the cost of the architecture and the probability of success of an attack. A solution of this optimization problem can be found by manually or automatically exploring all the possible sets of countermeasures or by more advanced optimization algorithms.

If the system is targeted by several root attacks, the definition of P can be refined by including the probability of success of the different root attacks weighted by their impact (for instance in terms of financial losses).

5 Discussions

5.1 Independent Attack Steps

As explained before, some attacks require several steps to be achieve. In other words, to perform a root or an intermediate attack A, an adversary may need to mount other q elementary attacks $A_1, ..., A_q$. For now, we have supposed that these elementary attacks are independent which means for instance that the resources needed to perform all the elementary attacks are the sum of the resources needed by each elementary attack.

However, two elementary attacks A_1 and A_2 can both require the same expensive piece of equipment (for instance an oscilloscope). In this case, the resources needed to perform the two attacks is not the sum (we do not require two oscilloscopes, one is sufficient). However, when two attacks are not independent, we have chosen to represent them as three independent attacks that all have to be performed to succeed, one (A_c) that contains the shared resources (for instance the oscilloscope) and the two other (A_1' and A_2') that embeds the resources specific for A_1 and A_2.

5.2 *SEQUENCE* vs *AND*

Sometimes, when several steps are required to perform an attack, these steps may need to be performed in a specific order. For this case, it could be interesting to define a *SEQUENCE* operator, in addition to the *AND* operator. However, the propagation rules of these two operators will be the same in our case.

5.3 A 2-Uple Attacker vs a 3-Uple One

Instead of describing an attacker by its resources R_M and its level of expertise E_M, we could have decomposed its resources into two parts: its financial resources F_M (which condition the hardware equipment it has access to) and the time window T_M it has to perform its attack. However, T_M and R_M are not always independent, for instance, an attacker may use money to buy extra equipment and save time.

Thus, we chose to represent time and financial resources in one parameter R (that includes the potential trade-off between financial resources and time) and define the attacker as a 2-uple (R, E). For the same reasons, an attack is defined as a 3-uple (l, R, E) and not a 4-uple (l, T, F, E).

6   Case Study

Let us consider the same system S that we have already described in Fig. 1 but with a more detailed ADT-SA.

Figure 7 represents an ADT-SA for an embedded system with a processor, an external RAM (not embedded inside the processor), a bus between the processor and the RAM. A small operating system and an application that performs cryptographic operations (such as encryption), using a secret key, runs on this system.

The attacker M wants to retrieve the confidential data manipulated by the system (in our case, the secret key manipulated by the application). The root attack A_{root} is *GetConfidentialData*. To perform this attack, the attacker M can either perform A_{bus} a bus probing attack that targets the component *Bus*, A_{SCA} a side-channel attack against the *Processor* during the manipulation of the secret key or A_{OS} a buffer overflow on the *Operating system* in order to take control of it and read confidential data directly from the memory.

We suppose that the required level of expertise to success an attack is represented by three values: *beginner, intermediate, expert*. We also arbitrarily fix the parameters of the three attacks:

- $R_{A_{bus}} = 40, E_{A_{bus}} = intermediate$
- $R_{A_{SCA}} = 50, E_{A_{SCA}} = intermediate$
- $R_{A_{OS}} = 90, E_{A_{OS}} = beginner$

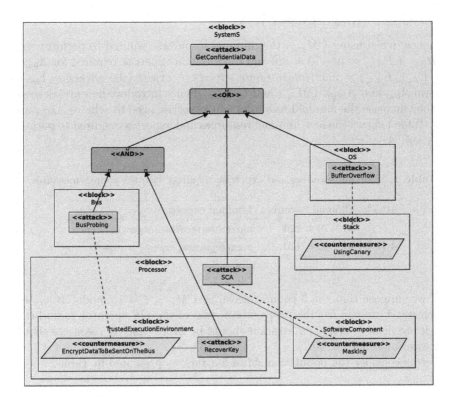

Fig. 7. Example of a more detailed ADT-SA for the system S

Three countermeasures are described on the ADT-SA:

- *EncryptDataToBeSentOnTheBus* (CM_{enc}) which consists in using a *Trusted Execution Enviroment TEE* that automatically encrypts data that are sent on the bus to the memory. The cost of data encryption on the bus is 1000 and the cost of TEE component is 2000. Thus, the cost of this countermeasure is $C_{enc} = 1000 + 2000 = 3000$. CM_{enc} adds a new attack step $N_{CM_{enc}} = \{A_{key}\}$. This new attack step (*RecoverKey* in the ADT-SA) consists in recovering the key used to encrypt the bus in order to decrypt data intercepted on the bus. For this new step: $R_{A_{key}} = 70$, $E_{A_{key}} = expert$.
- *Masking* (CM_{msk}) which protects against first order side-channel attacks. The cost of the countermeasure is $C_{msk} = 5000$ and it increases the difficulty of the attack A_{SCA} by requiring a second-order attack instead of a first-order one, so $I_{CM_{msk}} = \{(A_{SCA}, 70, expert)\}$.
- *UsingCanary* (CM_{OS}) which consists in using canary values to prevent buffer overflows on the stack from modifying the return address of

functions. The cost is $C_{cnry} = 200$ and it increases the difficulty of A_{OS} so $I_{CM_{cnry}} = \{(A_{OS}, 80, expert)\}$.

Hence, when using CM_{enc}, the minimal resources required to perform A_{bus} are $R_{A_{bus}} + R_{A_{key}} = 40 + 70 = 110$ and the minimal expertise required for A_{bus} is $\max(E_{A_{bus}}, E_{A_{key}}) = \max(intermediate, expert) = expert$ (the adversary has to perform A_{bus} and A_{key}). CM_{msk} and CM_{cnry} do not introduce new attack steps, but they increase the minimal resources and expertise need to achieve A_{SCA} and A_{OS}. Table 1 shows the new minimal resources and expertise required to perform A_{SCA} and A_{OS}:

Table 1. Minimal resources and expertise required with the countermeasures

Attack	Minimal resources	Minimal expertise
A_{SCA}	$50 + 70 = 120$	$\max(intermediate, expert) = expert$
A_{OS}	$90 + 80 = 170$	$\max(beginner, expert) = expert$

If we suppose that the 3 countermeasures CM_{enc}, CM_{msk} and CM_{cnry} are implemented, we can use the propagation rules of the OR operator to find that in order to achieve A_{root}, an attacker should be *expert* and its resources should be greater or equal to 110.

Let us consider the population $\mathcal{M}$ of 5 attackers presented in Table 2.

Table 2. Resources and expertise of each attacker in the population

Attacker	Resources	Expertise
M_1	130	*expert*
M_2	180	*intermediate*
M_3	60	*expert*
M_4	40	*expert*
M_5	70	*beginner*

We supposed that the system architect has fixed a budget of 14000 to build it. We supposed also that $C_{Total_{Comp}} = 6000$ (cost of the basic components of the system without the security equipments: OS, processor, stacks, bus, RAM). Thus, there are $14000 - 6000 = 8000$ for security investment i.e. use a TEE and implement CM_{enc}, CM_{msk} and CM_{cnry}. However, $C_{Total_{CM}|SA} = 5000 + 3000 + 200 = 8200$. Therefore, it could not deploy all of the countermeasures and it has to choose the set of countermeasures x that minimizes $P(x)$ for $\mathcal{M}$ and respect the constraint $C_{arch}(x) \leq 14000$.

If it chooses to implement CM_{enc} only, then the minimal resources and expertise needed to achieve A_{SCA} and A_{OS} will not increase. Thus, attackers with resources ≥ 50 and expertise $\geq$ *intermediate* will perform A_{SCA} and those whose resources ≥ 90 and expertise $\geq$ *beginner* will perform A_{OS} (*OR* operator). Hence, only attackers $M1$, $M2$ and $M3$ will successfully perform A_{root}, thus $P(\{CM_{enc}\}) = 3/5 = 0.6$ and $C_{arch}(x) = 6000 + 3000 \leq 14000$.

We can perform the same analysis for each set of countermeasures. Results are showed in Table 3 below.

Table 3. Costs and probability of success for each set of countermeasures

Countermeasures	Successful attackers	$C_{Total_{CM}|SA}$	C_{arch}	Prob. of success
$\{CM_{enc}\}$	$\{M_1, M_2, M_3\}$	3000	9000	$3/5 = 0.6$
$\{CM_{msk}\}$	$\{M_1, M_2, M_3, M_4\}$	5000	11000	$4/5 = 0.8$
$\{CM_{cnry}\}$	$\{M_1, M_2, M_3, M_4\}$	200	6200	$4/5 = 0.8$
$\{CM_{enc}, CM_{msk}\}$	$\{M_1, M_2\}$	8000	14000	$2/5 = 0.4$
$\{CM_{enc}, CM_{cnry}\}$	$\{M_1, M_2, M_3\}$	3200	9200	$3/5 = 0.6$
$\{CM_{msk}, CM_{cnry}\}$	$\{M_1, M_2, M_3, M_4\}$	5200	11200	$4/5 = 0.8$
$\{CM_{enc}, CM_{msk}, CM_{cnry}\}$	$\{M_1\}$	8200	14200	$1/5 = 0.2$

Hence, the set of countermeasures that solves our minimization problem is $x = \{CM_{enc}, CM_{msk}\}$ since $C_{arch}(x) = 14000$ and $P(x)$ is the minimal among the other values.

This example shows, on a small system, how our methodology can be used to find a good trade-off between the cost of an architecture and its security.

7   Conclusion

In this paper, we have presented a mechanism to help a system architect choose the right trade-off between the level of security of a system (that can be increased by adding countermeasures) and the total cost of the system. We improve attack-defense trees to describe the different attacks scenarios and how the countermeasures affect both the attack and the required hardware or software components. From the characterization of the elementary attack steps (in terms of resources and level of expertise required to successfully perform the attack), we use propagation rules to characterize root attacks. Given a population of malicious adversary, this leads to compute a probability of success of a root attack depending on the countermeasures that are implemented. From this point, the trade-off between security and cost can be expressed as an optimization problem (minimizing the probability of success of the adversary) with a constraint on the cost of the system.

We are currently working in two directions.

7.1 Automation

First, we are implementing our methodology in TTool [3] in order to help the system architect detail and refine attack-defense trees and make the trade-off between the cost of the architecture and the probability of success of the attacks. This is realized by automatically exploring the set of available countermeasures and computing the probability of success of attacks and architecture cost in the different configurations.

7.2 Attack-Defense Trees and Fault Trees

The next step is to take care of the safety aspect. A countermeasure increases the security of a system but it may degrade its safety by introducing new components that can fail. In addition, classical safety improvement techniques may degrade the security of the system by creating new attack scenarios. So there is a link between attack-defense trees and fault trees. We are working to formalize this link in order to allow the system architect to not only make a trade-off between security and cost but also take into account the safety.

Acknowledgments. This work is supported by the research chair *Connected Cars and Cyber Security* (C3S) [6] founded by Nokia, Renault, Thales, Valeo, Wavestone, Fondation Mines-Télécom and Télécom Paris.

References

1. A deep flaw in your car lets hackers shut down safety features. https://www.wired.com/story/car-hack-shut-down-safety-features/
2. Sysml-sec. http://sysml-sec.telecom-paristech.fr/
3. TTool. https://ttool.telecom-paristech.fr/
4. OMG Systems Modeling Language (OMG SysML), V1.0. Technical report, Object Management Group (2007). http://www.omg.org/spec/SysML/1.0/PDF
5. A survey on the usability and practical applications of graphical security models. Comput. Sci. Rev. **26**(C), 1–16 (2017). https://doi.org/10.1016/j.cosrev.2017.09.001
6. Research chair Connected Cars and Cyber Security (C3S) (2019). https://www.telecom-paristech.fr/c3s
7. Audinot, M., Pinchinat, S.: On the soundness of attack trees. In: Kordy, B., Ekstedt, M., Kim, D.S. (eds.) GraMSec 2016. LNCS, vol. 9987, pp. 25–38. Springer, Cham (2016). https://doi.org/10.1007/978-3-319-46263-9_2
8. Bistarelli, S., Dall'Aglio, M., Peretti, P.: Strategic games on defense trees. In: Dimitrakos, T., Martinelli, F., Ryan, P.Y.A., Schneider, S. (eds.) FAST 2006. LNCS, vol. 4691, pp. 1–15. Springer, Heidelberg (2007). https://doi.org/10.1007/978-3-540-75227-1_1
9. Edge, K., Dalton, G., Raines, R., Mills, R.: Using attack and protection trees to analyze threats and defenses to homeland security, pp. 1–7 (2006). https://doi.org/10.1109/MILCOM.2006.302512

10. Fraile, M., Ford, M., Gadyatskaya, O., Kumar, R., Stoelinga, M., Trujillo-Rasua, R.: Using attack-defense trees to analyze threats and countermeasures in an ATM: a case study. In: Horkoff, J., Jeusfeld, M.A., Persson, A. (eds.) PoEM 2016. LNBIP, vol. 267, pp. 326–334. Springer, Cham (2016). https://doi.org/10.1007/978-3-319-48393-1_24

11. Garro, A., Tundis, A.: A model-based method for system reliability analysis (2012)

12. Ji, X., Yu, H., Fan, G., Fu, W.: Attack-defense trees based cyber security analysis for CPSs. In: 2016 17th IEEE/ACIS International Conference on Software Engineering, Artificial Intelligence, Networking and Parallel/Distributed Computing (SNPD), pp. 693–698 (2016). https://doi.org/10.1109/SNPD.2016.7515980

13. Jürgenson, A., Willemson, J.: Computing exact outcomes of multi-parameter attack trees. In: Meersman, R., Tari, Z. (eds.) OTM 2008. LNCS, vol. 5332, pp. 1036–1051. Springer, Heidelberg (2008). https://doi.org/10.1007/978-3-540-88873-4_8

14. Jürgenson, A., Willemson, J.: Serial model for attack tree computations. In: Lee, D., Hong, S. (eds.) ICISC 2009. LNCS, vol. 5984, pp. 118–128. Springer, Heidelberg (2010). https://doi.org/10.1007/978-3-642-14423-3_9

15. Kordy, B., Kordy, P., Mauw, S., Schweitzer, P.: ADTool: security analysis with attack–defense trees. In: Joshi, K., Siegle, M., Stoelinga, M., D'Argenio, P.R. (eds.) QEST 2013. LNCS, vol. 8054, pp. 173–176. Springer, Heidelberg (2013). https://doi.org/10.1007/978-3-642-40196-1_15

16. Kordy, B., Mauw, S., Radomirović, S., Schweitzer, P.: Foundations of attack–defense trees. In: Degano, P., Etalle, S., Guttman, J. (eds.) FAST 2010. LNCS, vol. 6561, pp. 80–95. Springer, Heidelberg (2011). https://doi.org/10.1007/978-3-642-19751-2_6

17. Kordy, B., Piètre-cambacédès, L., Schweitzer, P.: Dag-based attack and defense modeling: Don't miss the forest for the attack trees. CoRR (2013)

18. Kordy, B., Wideł, W.: On quantitative analysis of attack–defense trees with repeated labels. In: Bauer, L., Küsters, R. (eds.) POST 2018. LNCS, vol. 10804, pp. 325–346. Springer, Cham (2018). https://doi.org/10.1007/978-3-319-89722-6_14

19. van Lamsweerde, A.: Elaborating security requirements by construction of intentional anti-models. In: Proceedings of the 26th International Conference on Software Engineering, ICSE 2004, pp. 148–157. IEEE Computer Society, Washington, DC, USA (2004). http://dl.acm.org/citation.cfm?id=998675.999421

20. Mauw, S., Oostdijk, M.: Foundations of attack trees. In: Won, D.H., Kim, S. (eds.) ICISC 2005. LNCS, vol. 3935, pp. 186–198. Springer, Heidelberg (2006). https://doi.org/10.1007/11734727_17

21. Saini, V., Duan, Q., Paruchuri, V.: Threat modeling using attack trees. J. Comput. Sci. Coll. **23**(4), 124–131 (2008). http://dl.acm.org/citation.cfm?id=1352079.1352100

22. Schneier, B.: Secrets & Lies: Digital Security in a Networked World, 1st edn. Wiley, New York (2000)

23. Steiner, M., Liggesmeyer, P.: Qualitative and quantitative analysis of CFTs taking security causes into account. In: Koornneef, F., van Gulijk, C. (eds.) SAFECOMP 2015. LNCS, vol. 9338, pp. 109–120. Springer, Cham (2015). https://doi.org/10.1007/978-3-319-24249-1_10

24. Zhou, S., Sun, Q., Jiao, J.: A safety modeling method based on SysML. In: 2014 10th International Conference on Reliability, Maintainability and Safety (ICRMS), pp. 1180–1185 (2014). https://doi.org/10.1109/ICRMS.2014.7107390

Security Analysis of IoT Systems Using Attack Trees

Delphine Beaulaton[1], Najah Ben Said[3], Ioana Cristescu[2(✉)], and Salah Sadou[1]

[1] University South Brittany, Irisa, Lorient, France
[2] Inria, Rennes, France
ioana-domnina.cristescu@inria.fr
[3] Thales SIX-GTS, Palaiseau, France

Abstract. Attack trees are graphical representations of the different scenarios that can lead to a security failure. In combination with model checking, attack trees are useful to quantitatively analyse the security of a system. Such analysis can help in the design phase of a system to decide how and where to modify the system in order to meet some security specifications.

In this paper we propose a security-based framework for modeling IoT systems where attack trees are defined alongside the model. A malicious entity uses the attack tree to exploit the vulnerabilities of the system. Successful attacks can be *rare events* in the system's execution, in which case they are hard to detect with usual model checking techniques. Hence, we use *importance splitting* as a statistical model checking technique for rare events. This technique requires a decomposition of an attack into sub parts, similarly to an attack tree. We argue that therefore, importance splitting is well suited, and benefits, from our modeling framework. We implemented our approach in a tool-set and verified its effectiveness by running a set of experiments over a real-word example.

Keywords: Attack tree · IoT · Rare events · Importance splitting

1 Introduction

The Internet of Things (IoT) is a rapidly emerging paradigm that provides a practical and easier way for users to manage and control a large variety of objects interacting over the Internet. However, IoT systems involve heterogeneous devices that are connected to a shared network and that carry critical tasks, and hence, are targets for malicious users. Vulnerabilities are discovered in opportunistic manner since security has mostly an ad-hoc treatment. Therefore, developing a systematic mechanism that considers security aspects at an early stage of system design helps detecting and preventing attacks.

Formal security analysis usually target systems with well-defined properties and specific implementation. For IoT systems, we have to find an appropriate abstraction level that is applicable to different implementations. Security issues

© Springer Nature Switzerland AG 2019
M. Albanese et al. (Eds.): GraMSec 2019, LNCS 11720, pp. 68–94, 2019.
https://doi.org/10.1007/978-3-030-36537-0_5

can occur at different levels in a system, for example in the computation nodes, in the communication protocols or at the storage level. *Attack trees* [12,15] are intuitive and practical formal methods to identify and analyze attacks. As their name suggests, attacks are modeled as trees, where the leaves represent elementary steps needed for the attack, and the root represents a *successful attack*. The internal nodes are of two types, indicating whether all the sub-goals (an AND node) or one of the sub-goals (an OR node) must be achieved in order to accomplish the main goal. Moreover, attack trees can express security issues of different nature in an uniform way. Hence, combining both formal analysis and attack trees helps to track and monitor the entire system in order to detect security breaches.

In this paper we present a framework to formally model IoT systems and analyse them using attack trees. In the formal modeling language we introduce, IoT systems are represented as a set of entities that communicate with each other if some verification on their identity holds. A malicious entity, called the *Attacker*, is explicitly represented in the system. The rest of the entities in the system may inadvertently help the Attacker by *leaking* their sensitive data. Equipped with the acquired knowledge the Attacker can then intrude the victim entities. Therefore, the system's vulnerabilities are also explicit in the model, and are represented by leaks.

We also propose a correct-by-construction transformation of an IoT model into a stochastic component-based model, called $\mathcal{SBIP}$ [1,3], for which an execution engine is developed and maintained. We can therefore execute our IoT model and run several verification and analysis tests, such as deadlock detection. Moreover, the attack tree provided with the model is transformed into a *monitor* that observes the interactions the Attacker has with the system and analyse when an attack is successful.

We then ask what is the probability of a successful attack given an IoT system and an attack tree. To respond to this question, we use two methods of statistical model checking (SMC) [3]: *Monte Carlo*, a standard SMC method, and *importance splitting* [8]. The Monte Carlo method consists of sampling the executions of an IoT system and computing the probability of an attack based on the number of executions for which the attack was successful. A successful attack can be considered a rare event if its probability value is in the range of 10^{-5} or 10^{-6}. For rare events, the Monte Carlo method can be problematic as it requires a large number of simulations for a correct estimate. We therefore use a second SMC method, developed for rare events, called *importance splitting* [8]. Importance splitting assumes that an execution leading to a rare event can be decomposed into several intermediate steps. Instead of executing a system until the rare event occurs, the execution is stopped after one of the intermediate steps is reached. The execution is restarted then from that step onward. Not only importance splitting can infer the probability of a rare event but it is also well suited for attack trees. The intermediate steps leading to a rare event are deduced from the nodes in the tree leading to a successful attack.

We implemented a tool chain to automate the analysis presented above. It consists of a compiler from our IoT modeling language into $\mathcal{S}$BIP. The execution engine of $\mathcal{S}$BIP is then used to produce simulations of the system, which are then fed to Plasma [4], a model checker that implements both Monte Carlo and importance splitting. Throughout our paper, we use a running example involving cyber-attacks on a Smart Hospital. The example is based on existing attacks carried against hospitals IT system as reported by TrapX [11] and ENISA [5].

The paper is structured as follows. Section 2 presents the IoT modeling language, Sect. 3 introduces attack trees and Sect. 4 presents $\mathcal{S}$BIP. The transformation from IoT to $\mathcal{S}$BIP is shown in Sect. 5. Section 6 presents the two SMC techniques. In Sect. 7 we validate our approach using some experiments on the running example. Related works that we are aware of, are summarised and compared with our work in Sect. 8. Lastly, Sect. 9 concludes.

2 Probabilistic IoT Models

The components of an IoT system, called *entities*, have each a *knowledge*, used to allow (or disallow) its interaction with the rest of the system. For instance, an entity can send an email to another entity only if it *knows* its email address. Or an user needs to *know* the url of a website in order to access it; we say that the url is part of the user's knowledge. For simplicity, we represent knowledge as a finite set of *values*.

Protocols are used at each interaction to verify the knowledge of the interacting entities. Each value is associated to a protocol. Two entities can communicate through a protocol if they have a common value for that protocol. We write C for a set of protocols, ranged over by c and *Val* for a set of values ranged over by v.

2.1 Processes and States

Each entity has a unique identifier, denoted by $e_1, \cdots e_n$ and a running process. The grammar of processes is defined in Fig. 1.

Processes are composed of *threads*, using the parallel composition operator. A thread can only do sequential computations. We write 0 for the inactive thread and A for the (recursive) definitions of threads.

The actions of a thread consists of sending and receiving values under an agreed upon protocol. We distinguish between "safe" interactions and the ones that can potentially lead to security issues, called *leak* and *collect*. A *leak* is a send action where there is no protocol governing the interaction and *collect* is its receive counterpart. Processes can also do silent moves, denoted by τ. Moreover, actions are equipped with a probability, denoted by $[n]a$, for an action a and a probability $n \in [0, 1]$. Threads can therefore do a probabilistic choice between actions, with the restriction that the sum of the probabilities of all available actions is 1. If there is only one available action, its probability is 1 and can be omitted.

$Process \quad P, Q \quad ::= T \parallel P \mid Q$

$Thread \quad T, U \quad ::= 0 \parallel A \parallel \sum_{i \in I} [n_i] a_i.T_i \text{ where } n_i \in (0, 1] \text{ and } \sum_{i \in I} n_i = 1$

$Action \quad a, b \quad ::= e \xrightarrow{c}_{v} e' \quad (\text{SEND}) \parallel e \xleftarrow{c}_{v} e' \quad (\text{RECEIVE}) \parallel$

$\qquad\qquad\qquad e \xrightarrow{}_{v} e' \quad (\text{LEAK}) \parallel e \leftarrow e' \quad (\text{COLLECT}) \parallel$

$\qquad\qquad\qquad \tau \quad (\text{INTERNAL})$

$Definition \quad A \overset{\text{def}}{=} T$

$\qquad State \quad s \quad ::= \emptyset \parallel \langle P, k \rangle \parallel s \mid s.$

Fig. 1. Syntax of the probabilistic IoT-calculus

A *knowledge* function $K : E \times C \to \mathcal{P}(Val)$ associates a set of values to each entity and protocol. For simplicity we write k_i^c for the knowledge of entity i under protocol c. The function protocol $: Val \to C$ associates each value to a protocol.

Each entity, has at any state of its computation, a running process P and a knowledge k. The grammar for states is included in Fig. 1. The global state of a IoT system consists of the parallel composition of all entities states $s_1 \mid \cdots \mid s_n$, where s_i is the current state of the entity e_i.

2.2 Operational Semantics

We define $\equiv_P \subseteq P \times P$ to be the smallest congruence on processes which includes the associativity and the commutativity for $+$ and $\mid$; the identity element 0 for $\mid$ and the unfolding law for definitions: $A \equiv_P T$ if $A \overset{\text{def}}{=} T$. We also introduce a congruence relation on states $\equiv_s \subseteq s \times s$ which includes the associativity and the commutativity for $\mid$, the identity element $\emptyset$ and which generalizes the congruence on processes: if $P \equiv_P Q$ then $\langle P, k \rangle \equiv_s \langle Q, k \rangle$.

The operational semantics of Fig. 2, defines a transition system (S, T, L, s_0) where we write S for the set of states, with s_0 the initial state, $L \subseteq \{\tau\} \cup (\{SR, LC\} \times Val)$ for a set of labels, ranged over by l, and $T \subseteq S \times [0, 1] \times L \times S$ for a set of transitions, where each transition is decorated by a probability and by a label. A transition can either be internal, labeled by τ, or it can be an interaction between two entities exchanging a value.

In our semantics, a probabilistic choice is always resolved locally, using the CHOICE rule. A transition derived by the CHOICE rule is considered internal and is labeled τ. A process can also do internal transitions using rule INTERNAL. Rule SENDRECEIVE defines the interaction between two components e_1 and e_2. The interaction is allowed if the sender and the receiver share some common values under the protocol c. After the interaction, the receiver's knowledge is updated by adding the received value under the corresponding protocol. A LEAKCOLLECT interaction proceeds similarly, except that there are no checks on the knowledge

CHOICE   INTERNAL

$$\langle \sum_{i \in I} [n_i] a_i.T_i, k \rangle \xrightarrow[\tau]{[n_i]} \langle a_i.T_i, k \rangle \qquad \langle \tau.P, k \rangle \xrightarrow[\tau]{[1]} \langle P, k \rangle$$

SENDRECEIVE

$$\frac{\exists v \in k_1^c \text{ s.t. } v \in k_2^c \qquad c' = \mathsf{protocol}(v')}{\langle e_1 \xrightarrow[v']{c} e_2.P_1, k_1 \rangle | \langle e_2 \xleftarrow{c} e_1.P_2, k_2 \rangle \xrightarrow[SR:v']{[1]} \langle P_1, k_1 \rangle | \langle P_2, k_2^{c'} \uplus \{v'\} \rangle}$$

LEAKCOLLECT

$$\frac{c' = \mathsf{protocol}(v')}{\langle e_1 \xrightarrow{v'} e_2.P_1, k_1 \rangle | \langle e_2 \leftarrow e_1.P_2, k_2 \rangle \xrightarrow[LC:v']{[1]} \langle P_1, k_1 \rangle | \langle P_2, k_2^{c'} \uplus \{v'\} \rangle}$$

PARPROC   CONGRUENCE

$$\frac{\langle P_i, k_i \rangle | \langle P_j, k_j \rangle \xrightarrow[l]{[n]} \langle P_i', k_i' \rangle | \langle P_j', k_j' \rangle}{\langle P_i \mid Q_i, k_i \rangle | \langle P_j \mid Q_j, k_j \rangle \xrightarrow[l]{[n]} \langle P_i' \mid Q_i, k_i' \rangle | \langle P_j' \mid Q_j, k_j' \rangle} \qquad \frac{s \equiv_s t \xrightarrow[l]{[n]} s' \equiv_s t'}{s \xrightarrow[l]{[n]} s'}$$

PARSTATE_TAU $\qquad \dfrac{\langle P, k \rangle \xrightarrow[\tau]{[n]} \langle P', k' \rangle \qquad \mathsf{count}_\tau(\langle P, k \rangle | s) = m}{\langle P, k \rangle | s \xrightarrow[\tau]{[1/m \cdot n]} \langle P', k' \rangle | s}$

PARSTATE_INTERACTION

$$\frac{\langle P_i, k_i \rangle | \langle P_j, k_j \rangle \xrightarrow[l]{[1]} \langle P_i', k_i' \rangle | \langle P_j', k_j' \rangle \qquad \mathsf{count}_{SR,LC}(\langle P_i, k_i \rangle | \langle P_j, k_j \rangle | s) = m \qquad \mathsf{count}_\tau(\langle P_i, k_i \rangle | \langle P_j, k_j \rangle | s) = 0}{\langle P_i, k_i \rangle | \langle P_j, k_j \rangle | s \xrightarrow[l]{[1/m]} \langle P_i', k_i' \rangle | \langle P_j', k_j' \rangle | s}$$

Fig. 2. The operational semantics of an IoT system

of the two components. Rules CONGRUENCE and PARPROC allows one to use congruence and parallel composition on states to derive transitions.

The rules for the global states, PARSTATE_TAU and PARSTATE_INTERACTION, give priority to the internal transitions over the binary interactions. Moreover, in each case, we choose a global transition from several local ones using an uniform distribution. We rely on two auxiliary functions, count_τ and $\mathsf{count}_{SR,LC}$ that count the number of local transitions with labels τ and labels SR, LC, respectively.

Definition 1. *Given an IoT model (S, T, L, s_0) with an initial state s_0, an execution is a sequence of transitions in T, $\sigma = \{s_i \xrightarrow[l_i]{[n_i]} s_i'\}_{0 \leq i \leq k}$, such that $s_0 = s$ and $\forall i \geq 1$, $s_{i-1}' = s_i$. The probability of a transition $s \xrightarrow[l]{[n]} s'$ is n and the probability of an execution σ is $\prod_{0 \leq i \leq k} n_i$.*

Example 1. Let us now introduce our running example, the Smart Hospital system. Let $E = \{A(ttacker), H(ospital), E(mployee)\}$ be three entities which communicate with each other using the protocols $C = \{url, message, mail, phone\}$.

We introduce the following actions:

$$\text{AH} = A \xrightarrow[\text{getSensitiveData}]{\text{url}} H \qquad\qquad \text{leakEmail} = H \xrightarrow[\text{emailEmployee}]{} A$$

$$\text{HA} = H \xleftarrow{\text{url}} A \qquad\qquad\qquad \text{leakPhone} = H \xrightarrow[\text{phoneEmployee}]{} A$$

$$\text{AE\_mail} = A \xrightarrow[\text{getCredential}]{\text{mail}} E \qquad\quad \text{leakIssue} = H \xrightarrow[\text{issueEmployee}]{} A$$

$$\text{EA\_mail} = E \xleftarrow{\text{mail}} A \qquad\quad \text{leakCredentials} = E \xrightarrow[\text{credEmployee}]{} A$$

$$\text{AE\_phone} = A \xrightarrow[\text{getCredential}]{\text{phone}} E$$

$$\text{EA\_phone} = E \xleftarrow{\text{phone}} A$$

and process definitions:

Attacker $=$ AChoice $\mid$ collectH $\mid$ collectE

AChoice $= [0.4]$AH.AChoice $+ [0.3]$AE_mail.AChoice $+ [0.3]$AE_phone.AChoice

collectH $= A \leftarrow H$.collectH

collectE $= A \leftarrow E$.collectE

Hospital $=$ HA.$([n_1]$leakPhone.Hospital $+ [n_2]$leakIssue.Hospital$+$
$\qquad\qquad [n_3]$leakEmail.Hospital $+ [n_4]\tau$.Hospital)

Employee $= [m_1]$EA_mail.EChoice $+ [m_2]$EA_phone.EChoice

EChoice $= [m_3]$leakCredentials.Employee $+ [m_4]\tau$.Employee

A has Attacker as initial process and similarly for H and E. Their initial knowledge is:

$k_A = \{url = \{\text{urlHospital}\}, message = \{\text{getSensitiveData, getCredentials}\}\}$

$k_H = \{url = \{\text{urlHospital}\}, mail = \{\text{emailEmployee}\}, phone = \{\text{phoneEmployee}\}\}$

$k_E = \{mail = \{\text{emailEmployee}\}, phone = \{\text{phoneEmployee}\}\}$

where the missing protocols are initially the emptysets.

Let us now consider the transitions below. The Attacker starts by choosing to contact the Hospital using the internal transition (1), after which the two entities can communicate in transition (2). At this point the Hospital can either leak some sensitive information (*emailEmployee*, *phoneEmployee* or *issueEmployee*) or it can do an internal transition. Transitions (3) − (4) represent the scenario where *emailEmployee* is leaked. The knowledge of the attacker is augmented with the leaked data: $k_A^{mail} \cup \{\text{emailEmployee}\}$ and the attacker can now communicate with the employee using the email protocol.

$$\langle \text{Attacker}, k_A \rangle \mid \langle \text{Hospital}, k_H \rangle \xrightarrow[\tau]{[0.4]} \tag{1}$$

$$\langle \text{AH.AChoice} \mid \text{collectH} \mid \text{collectE}, k_A \rangle \mid \langle \text{Hospital}, k_H \rangle \xrightarrow[\text{SR:getSensitiveData}]{[1]} \tag{2}$$

$$\langle \text{Attacker}, k_A \rangle \mid \langle [n_3]\text{leakEmail.Hospital} + \cdots, k_H' \rangle \xrightarrow[\tau]{[n_3]} \tag{3}$$

$$\langle \text{Attacker}, k_A \rangle \mid \langle \text{leakEmail.Hospital}, k_H' \rangle \xrightarrow[\text{LC:emailEmployee}]{[1]} \tag{4}$$

$$\langle \text{Attacker}, k_A' \rangle \mid \langle \text{Hospital}, k_H' \rangle$$

Suppose that after transition (2), the Hospital does not leak any information:

$$\langle \text{Attacker}, k_A \rangle \mid \langle [n_3]\text{leakEmail.Hospital} + \cdots, k_H' \rangle \xrightarrow[\tau]{[n_4]}$$

$$\langle \text{Attacker}, k_A \rangle \mid \langle \text{Hospital}, k_H' \rangle.$$

Then the Attacker cannot communicate with the Employee. If the Attacker tries to communicate without knowing the *emailEmployee*, the system deadlocks. A sequence of transitions ending with a deadlock represents an unsuccessful attack.

3   Attack Trees

In this section we formally introduce attack trees and we show how attack trees can be used to monitor the execution of an IoT system.

Figure 3 shows an attack tree for the Smart Hospital in Example 1. The root of the tree is the main goal of the Attacker, which is getting the Employee's credentials. The nodes represent the possible attacks. For example the *qui pro quo attack* consists of contacting the Employee and posing as a technician. For this attack, node *get information* has to happen as well. It requires for the Attacker to get the Employee's phone number and technical issue, which both are leaked by the Hospital. The second possibility is a *phishing attack*. It consists of the Attacker contacting the Hospital and then the Hospital leaking the Employee's email. If either of the two attacks succeed, the Attacker can then try to get the Employee's credentials.

Formally, the leaves of the tree corresponds to *events* happening in the system. An event in an IoT system consists of the exchange of a value between some entities. For example, the node *Employee leaks credentials* stands for the pair *(LC, "credEmployee")*, meaning that the value *credEmployee* has been leaked, through a LeakCollect interaction.

Definition 2. *Let $\Delta \subseteq \{SR, LC\} \times Val$ be a set of events. An* attack tree t *is a term constructed recursively from the set Δ using the operators $\vee$ and $\wedge$.*

An attack tree can also be seen as a Boolean expression, by associating to every event $e \in \Delta$ a Boolean variable v_e.

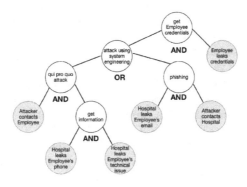

Fig. 3. An attack tree for the Smart Hospital

Definition 3. *Let t be an attack tree. The semantics of t, denoted $[\![t]\!]$, consists of a Boolean expression defined by recursion on t as follows:*

- *if $e \in \Delta$ then $[\![t]\!] = v_e$;*
- *if $t = t_1 \wedge t_2$ then $[\![t]\!] = [\![t_1]\!] \wedge [\![t_2]\!]$;*
- *if $t = t_1 \vee t_2$ then $[\![t]\!] = [\![t_1]\!] \vee [\![t_2]\!]$.*

Let $\mathbf{X} : \Delta \rightarrow \{true, false\}$ be a valuation for Δ, then the semantics of t w.r.t. $\mathbf{X}$, denoted $[\![t]\!](\mathbf{X}) \in \{true, false\}$, consists in evaluating the associated Boolean formula [9].

In order to assess whether an attack was successful, we can use an attack tree to monitor the executions of an IoT system. Given an execution trace σ, its corresponding valuation $\mathbf{X}(\sigma)$ sets v_e to *true* if the event e occurred in σ. If $[\![t]\!](\mathbf{X}(\sigma))$ is true then the execution σ is a successful attack of t.

4   SBIP: A Stochastic Component Based Model

$\mathcal{S}$BIP [1,3] is a stochastic, component based framework that allows modeling hierarchical systems from *atomic* components. We introduce $\mathcal{S}$BIP in four steps: we start with some preliminary notations; then we introduce the syntax of the atomic components; next its semantics; and lastly we explain how to compose atomic components into hierarchical systems.

4.1   Preliminaries

Let V be a set of variables, and for each variable $v_j \in V$, let D_j be its data domain, denoted by $v_j : D_j$. A *valuation* for the variables in V is a function $\mathbf{X} : V \rightarrow \cup_j D_j$ which assigns values to variables. We denote $\mathbf{X}(v)$ the valuation of the variable $v \in V$.

Let $\mathbb{E}$ be a set of operators. We denote by $\mathbb{E}[V]$ the set of expressions constructed from a set of variables V and operators. A function $f(V)$ is then just an expression in $\mathbb{E}[V]$. We denote $\mathbf{X}(e)$ the valuation of the expression $e \in \mathbb{E}[V]$ according to the valuation of the variables in V.

We write $v := e$ for an *assignment*, or *update* of v, and write $\mathbb{A}[V]$ for a set of assignments for the variables in V.

We distinguish between two types of variables: the deterministic variables and the *random* variables, used for encoding the stochastic behavior. A random variable v is associated with a probability distribution μ over its valuation domain D, denoted as $v \sim \mu$, where $\mu : D \to [0,1]$ and $\sum_{x \in D} \mu(x) = 1$.

Lastly, we denote $(f \circ g)(x) = f(g(x))$ the composition of functions. If the two functions have disjoint domains, we write $f \sqcup g$ for their disjoint composition. We use $\cup$ for set union and $\uplus$ for disjoint union.

4.2   Stochastic Atomic Components

$\mathcal{S}$BIP components are 1-safe Petri-Nets equipped with (i) ports that allow the component to communicate with other components; and (ii) variables that can be read and updated during communications.

Definition 4. *A* stochastic atomic component *consists of the tuple* $\mathcal{B} = (P, V, N)$*, where*

- *P is a set of communication ports.*
- *$V = V^d \uplus V^p$, with $V^d = \{v_1, \ldots, v_n\}$ a set of deterministic variables and $V^p = \{v_1^p, \ldots, v_m^p\}$ a set of random variables with an associated distribution $v_i^p \sim \mu_i$.*
- *$N = (L, L_0, T)$ is a Petri-Net[1] where L is a set of places and $L_0 \subseteq L$ are the initial places. T is a set of transitions $t = (^\bullet t, \langle p, g, f \rangle, t^\bullet)$ where $^\bullet t$ (resp. $t^\bullet$) is the set of input (resp. output) places of t. Transitions are labeled by the triple $\langle p, g, f \rangle$ where $p \in P$ is a port, $g \in \mathbb{E}[V]$ is a guard and $f = (f^d, R^p)$ is an update function, such that $f^d = \{v := f(V) \mid v \in V^d\} \in \mathbb{A}[V]$ is a set of functions that update the deterministic variables and $R^p \subseteq V^p$ is a subset of random variables to be updated.*

We sometimes write p_t, g_t and f_t^d, R_t^p for the label of t. We define *markings* as the set of functions $m : L \to \{0, 1\}$. Given two markings m_1, m_2 we define inclusion $m_1 \leq m_2$ iff for all $l \in L$, $m_1(l) \leq m_2(l)$. Also, we define addition $m_1 + m_2$ as the marking m_{12} such that, for all $l \in L$, $m_{12}(l) = m_1(l) + m_2(l)$.

A *priority order* on a set of ports is a partial order, where each element $p < p'$ of the order is called a *priority*. Whenever the system has a choice between the two interactions on two ports p or p', the interaction on p' is chosen.

---

[1] N is equivalent to the extended 1-safe Petri-Net (L, L_0, T, F) where $F = \{(l, t) \mid l \in {}^\bullet t\} \cup \{(t, l) \mid l \in t^\bullet\}$ is the token flow relation and can be deduced from T.

4.3 Semantics of Stochastic Atomic Components

The semantics of a $\mathcal{S}$BIP component $\mathcal{B} = (P, V, N)$ consists of a transition system $\mathcal{M}$, where the states are of the form $(m, \mathbf{X})$, for m a marking of N and $\mathbf{X}$ a valuation of V.

The random variables engender a probabilistic behavior over transitions of $\mathcal{M}$. Let us consider an atomic component $\mathcal{B}$ in Fig. 4a that has a transition going from place l_1 to place l_2 using port p, with a guard that is always true, and which updates a random variable v with the valuation domain D and distribution μ. Assuming the initial value of v is x_0, when executing $\mathcal{B}$, there will be several possible transitions, shown in Fig. 4b, from state $(\{l_1\}, x_0)$ to states $(\{l_2\}, x_i)$ for all $x_i \in D$. The probabilities of these transitions is given by μ. Since the random variables are independent, when several random variables are updated, the resulting distribution on transitions is the product of the distributions associated to each variable.

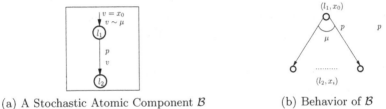

(a) A Stochastic Atomic Component $\mathcal{B}$ (b) Behavior of $\mathcal{B}$

Fig. 4. Example of a stochastic atomic component $\mathcal{B}$ and its behavior.

Atomic components with random variables lead to behaviors that combine both stochastic and non-deterministic aspects. A transition is possible if a communication is ready on its associated port. At any given state, several ports can be ready for a communication, and thus several transitions can be enabled, regardless of whether they are associated or not with random variables. Non-determinism is always resolved in $\mathcal{S}$BIP to a probabilistic choice on an uniform distribution. To formally state this, we denote with Enabled$(m; \mathbf{X})$ the set of transitions in T that are enabled by m for a valuation $\mathbf{X}$: Enabled$(m; \mathbf{X}) = \{t \in T \mid {}^\bullet t \leq m \text{ and } \mathbf{X}(g_t) \text{ is true}\}^2$.

Definition 5. *The semantics of a stochastic component* $\mathcal{B} = (P, V, (L, L_0, T))$ *with* $\mathbf{X}_{\mathbf{init}}$ *an initial valuation, is defined as a probabilistic transition system* $\mathcal{M} = \langle Q, \pi, P, q_0 \rangle$, *where:*

- Q *is a set of states of the form* $(m, \mathbf{X})$; $q_0 = (m_0, \mathbf{X}_{\mathbf{init}})$ *is the initial state where* m_0 *is the marking associated to* L_0, *i.e.* $m_0(l) = 1 \iff l \in L_0$ *and* 0 *otherwise;*

---

[2] Remark that the cardinality of Enabled$(m; X)$ can be greater than one.

– $\pi \subseteq Q \times P \times Q$ is a set of transitions defined by the following rule:

$$\frac{t \in T \quad {}^{\bullet}t \leq m \quad m' = m - {}^{\bullet}t + t^{\bullet} \quad \mathbf{X}(g_t) = true}{\mathbf{X}' = [v^d := \mathbf{X}(f_t^d), v^p := random(\mu)] \quad v^d \in V^d \quad v^p \in R_t^p, v^p \sim \mu}{(m, \mathbf{X}) \xrightarrow{p_t} (m', \mathbf{X}')}$$

Lastly, we defined the probability of a transition as follows:

$$\mathbb{P}(q \xrightarrow{p} q') = \frac{1}{|\mathsf{Enabled}(m; \mathbf{X})|} \cdot \prod_{v_i \in R^p, v_i \sim \mu_i} \mu_i(X'(v_i)).$$

In the definition above we say that the state $(m', \mathbf{X}')$ is a successor of state $(m, \mathbf{X})$, if t is a transition of T enabled by the marking m, the guard g_t evaluates to *true* and the new valuation $\mathbf{X}'$ on the variables $V^d \cup V^p$ is obtained by applying f_t^d on the deterministic variables V^d and updating the random variables in R_t^p. The probability of a transition is computed by first selecting a transition with an uniform distribution from the set of enabled transitions; and then, selecting the next state according to the distributions attached to the random variables.

4.4 Composition of Stochastic Components

Definition 6. *An* interaction $\gamma = (P, G, F)$ *on a set of components* $\mathcal{B}_i = (P_i, V_i, N_i)$, *for* $i \in I$, *where* I *is set of indexes, consists of:*

– $P = \{p_i \mid p_i \in P_i, i \in I\}$ *is a disjoint set of ports containing exactly one port from each* $\mathcal{B}_i$, $i \in I$;
– G *is a global guard defined on* $V_\gamma = \cup_{i \in I} V_i$;
– $F = \{v := F(V_\gamma) \mid v \in \cup_{i \in I} V_i^d\}$ *is a global update function used to exchange values between components.*

For n atomic components and for Γ a set of interactions, we write $\Gamma(\mathcal{B}_1, \ldots, \mathcal{B}_n)$ their composition into a stochastic component. Intuitively the local transitions of the atomic components synchronise to produce global transitions using the interactions.

Definition 7. *Let* Γ *be a set of interactions defined on* n *components* $\mathcal{B}_i = (P_i, V_i, N_i)$, *with* $N_i = (L_i, L_{0,i}, T_i)$ *for* $i \leq n$. *The composition of the* n *components, denoted as* $\Gamma(\mathcal{B}_1, \ldots, \mathcal{B}_n)$, *is a stochastic component* $\mathcal{B} = (\Gamma, V, N)$, *with* $N = (L, L_0, T)$, *defined as follows:*

– $V = \cup_{i \leq n} V_i$;
– $L = \cup_{i \leq n} L_i$ *with* $L_0 = \cup_{i \leq n} L_{0,i}$;
– $T = \{({}^{\bullet}T_\gamma, \langle \gamma, g, f \rangle, T_\gamma^{\bullet}) \mid \gamma \in \Gamma\}$, *where* $T_\gamma = \{t_i \mid p_i \in P_\gamma\}$ *is the set of transitions that synchronize on the interaction* $\gamma \in \Gamma$. *Then* ${}^{\bullet}T_\gamma = \{l \mid l \in {}^{\bullet}t_i, t_i \in T_\gamma\}$ *and* $T_\gamma^{\bullet} = \{l \mid l \in t_i^{\bullet}, t_i \in T_\gamma\}$. *Each transition is labeled by the triple* $\langle \gamma, g, f \rangle$ *where* $g = G_\gamma \wedge (\bigwedge_{t_i \in T_\gamma} g_{t_i})$ *and* $f = F_\gamma \circ (\sqcup_{t_i \in T_\gamma} f_{t_i})$ *consists of the composition of all* f_{t_i} *with* F_γ.

Assembling stochastic atomic components produces a stochastic atomic component, and thus its semantics is given by Definition 5.

We use a priority order, denoted $\ll$, which gives priority to the internal transitions over the binary interactions. We write then $\langle\ll\rangle(\Gamma(\mathcal{B}_1,\ldots,\mathcal{B}_n))$ for a $\mathcal{S}$BIP system.

5 Transformation from IoT to $\mathcal{S}$BIP

We now show how to transform an IoT system to $\mathcal{S}$BIP. Entities of an IoT model become atomic components and their communications is represented as interactions.

An entity can have several threads running, all sharing the same knowledge. To model this we encode each thread of a process into a Petri Net (Definition 8). The encoding of a process is then the union of the several Petri Nets, which all have a common set of variables, guards and update functions (Definition 9).

The deterministic variables are used to model the entity's knowledge. The random variables, similarly to [3], encode the probabilities associated to actions in a summation process.

We use labeling functions on places and on the random variables, denoted by ℓ. The labels are the threads of the original IoT system. Moreover we identify places that have congruent labels, i.e. $l_1 \equiv_L l_2 \iff \ell(l_1) \equiv_P \ell(l_2)$. We write l_T and v_T when $\ell(l) = T$ and $\ell(v) = T$, respectively.

Definition 8. *For a thread T, let* Definitions *and* Actions *be the sets of thread definitions and of actions, respectively, used recursively in T. We define the transformation of T to be the atomic component* (Actions, $V^d \uplus V^p, (\mathcal{L}, \mathcal{L}_0, \mathcal{T}))$ *with:*

- $V^d = \{v_c \mid c \text{ is a protocol used in } T\}$;
- $V^p = [\![T]\!]_v \cup \{[\![U]\!]_v \mid A \stackrel{def}{=} U \text{ and } A \in \text{Definitions}\}$;
- $\mathcal{L} = \left([\![T]\!]_s \cup \{[\![U]\!]_s \mid A \stackrel{def}{=} U \text{ and } A \in \text{Definitions}\}\right)_{\equiv_L}$ *is a set of places partitioned in equivalence classes by the $\equiv_L$ relation, with $\mathcal{L}_0 = \{l_T\}$;*
- $\mathcal{T} = [\![T]\!]_t \cup \{[\![U]\!]_t \mid A \stackrel{def}{=} U \text{ and } A \in \text{Definitions}\}$.

In our transformation we use the functions $[\![\cdot]\!]_v$, $[\![\cdot]\!]_s$ and $[\![\cdot]\!]_t$ to transform a thread into a set of places, random variables and transitions. The functions are formally defined in Fig. 5. Intuitively, for each possible continuation of T we introduce a new place, and we use the labeling function on places to keep track of the correspondence between places and threads. Also whenever T is of the form $\sum_{i\in I}[n_i]a_i.T_i$ we introduce a new place, denoted $l_T^\star$. This additional place is where the choice between the different branches of the sum is made. The random variables are defined using the function $[\![\cdot]\!]_v$. Whenever T is of the form $\sum_{i\in I}[n_i]a_i.T_i$ we introduce a new random variable v_T. The valuation domain D for v_T is the set of states associated to the possible continuations i.e. $D = \{a_i.T_i\}_{i\in I}$. The probability distribution of v_T is defined by the probabilities

n_i i.e. $\mu(a_i.T_i) = n_i$. Lastly, $\llbracket T \rrbracket_s$ defines the transitions. The guards are only used when making a probabilistic choice: Suppose we are currently running thread T and we wish to go from state $l_T^\star$ to a state l_{T_i}. The guard then checks that the value of the random variable v_T is updated to $a_i.T_i$. For the rest of transitions, the guard is the constant $true$.

$$\llbracket \sum_{i \in I}[n_i]a_i.T_i \rrbracket_v = \bigcup_{i \in I} \llbracket a_i.T_i \rrbracket_v \cup \{v_T \mid v_T \sim \mu \text{ s.t. } \mu(a_i.T_i) = n_i, \forall i \in I\}$$

$$\text{where } T = \sum_{i \in I}[n_i]a_i.T_i \text{ and } |I| > 1$$

$$\llbracket a.T \rrbracket_v = \llbracket T \rrbracket_v$$

$$\llbracket A \rrbracket_v = \llbracket 0 \rrbracket_v = \emptyset$$

$$\llbracket \sum_{i \in I}[n_i]a_i.T_i \rrbracket_s = \bigcup_{i \in I} \llbracket T_i \rrbracket_s \cup \{l_T, l_T^\star\}, \text{ where } T = \sum_{i \in I}[n_i]a_i.T_i \text{ and } |I| > 1$$

$$\llbracket a.T \rrbracket_s = \llbracket T \rrbracket_s \cup \{l_{a.T}\}$$

$$\llbracket A \rrbracket_s = \{l_A\}$$

$$\llbracket 0 \rrbracket_s = \{l_0\}$$

$$\llbracket \sum_{i \in I}[n_i]a_i.T_i \rrbracket_t = \bigcup_{i \in I} \left((\{l_T^\star\}, \langle a_i, g = (v_T == a_i.T_i), f \rangle, \{l_{T_i}\}) \cup \llbracket T_i \rrbracket_t \right)$$

$$\cup (\{l_T\}, \langle \tau, \text{true}, f^\star \rangle, \{l_T^\star\}) \text{ where } T = \sum_{i \in I}[n_i]a_i.T_i \text{ and } |I| > 1$$

$$\llbracket a.T \rrbracket_t = (\{l_{a.T}\}, \langle a, \text{true}, f \rangle, \{l_T\}) \cup \llbracket T \rrbracket_t$$

$$\llbracket 0 \rrbracket_t = \llbracket A \rrbracket_t = \emptyset$$

where $f = \{v := v \mid v \in V^d\}$ and $R^p = \emptyset$ and $f^\star$ defined as f but with $R^p = \{v_T\}$.

Fig. 5. The functions used in the transformation in Definition 8

Definition 9. *Let e be an entity in an IoT system with the initial state $s_0 = \langle P, k \rangle$ and $P = T_1 \mid \cdots \mid T_m$. Let $(P^j, V^j, (\mathcal{L}^j, \mathcal{L}_0^j, \mathcal{T}^j))$ be the atomic components obtained from each T_j, $j \leq m$.*

We define the transformation of e as the atomic component $\mathcal{B}_e = (P, V, N)$ with $P = \cup_{j \leq m} P_j$, $V^d = \cup_{j \leq m} V_j^d$, $V^p = \uplus_{j \leq m} V_j^p$ and with $N = (\uplus_{j \leq m} \mathcal{L}^j, \uplus_{j \leq m} \mathcal{L}_0^j, \uplus_{j \leq m} \mathcal{T}^j)$. We also define the initial valuation $\mathbf{X_{init}}(v_c) = k(c)$ where for each protocol c we initialize the variable v_c to the set of values $k(c)$.

Note that we are using set union for ports and variables as the different threads of an entity share their ports and knowledge. However, in order to clearly separate the behaviour of the different threads, we use disjoint union when combining the Petri Nets of the different threads. This is allowed because the different threads in an entity cannot interact with each other, but only with other entities.

Communications between two entities e_1 and e_2 in the IoT language are transformed into a set of guarded interactions between components $\mathcal{B}_{e_1}$ and $\mathcal{B}_{e_2}$.

Definition 10. *Let* $\mathcal{B}_{e_i} = (P_i, V_i, N_i)$ *be the transformation of an IoT system with n entities e_i and with the initial state s_0. For all $a \in$ Actions, if there exists $a' \in$ Actions such that*

- *either* $a = e_1 \xrightarrow[v]{c} e_2$ *and* $a' = e_2 \xleftarrow{c} e_1$,
- *or* $a = e_1 \xrightarrow[v]{} e_2$ *and* $a' = e_2 \leftarrow e_1$

then we define an interaction $\gamma = (\{a, a'\}, G, F)$ *where*

- *if* $a = e_1 \xrightarrow[v]{c} e_2$ *then* $G = (\exists x \in v_c^1$ *such that* $x \in v_c^2)$ *for* $v_c^1 \in V_1^d$, $v_c^2 \in V_2^d$; *otherwise* $G = true$;
- $F = \{v_{c'}^2 := v_{c'}^2 \cup \{v\} \mid \mathsf{protocol}(v) = c', v_{c'}^2 \in V_2^d\}$

and where V_1^d, V_2^d *are the deterministic variables of* $\mathcal{B}_{e_1}$ *and* $\mathcal{B}_{e_2}$, *respectively.*

We also define the interaction $(\{\tau\}, true, F)$, *where* $F = \{v := v \mid v \in V^d\}$, *for every component* $\mathcal{B}_e = (P, V, N)$.

Note that interactions are only defined for consistent pairs of *SendReceive* and *LeakCollect* reflecting the rules in Fig. 2.

Given an IoT system with n entities and an initial state s_0 let us write Γ for the set of interactions of Definition 10. Recall that $\ll$ is the priority order of Sect. 4.4 and that $\Gamma(\mathcal{B}_1, \ldots, \mathcal{B}_n)$ is the composition of the n entities, as in Definition 7. Then $\langle\ll\rangle(\Gamma(\mathcal{B}_1, \ldots, \mathcal{B}_n))$ is the resulting $\mathcal{S}$BIP system. We have everything in place to show a correspondence between the *semantics* of the IoT system and of its $\mathcal{S}$BIP transformation.

Theorem 1. *Let* (S, L, T, s_0) *be an IoT system where* $s_0 = \langle P_1, k_1 \rangle \mid \cdots \langle P_n, k_n \rangle$ *and where* $\langle P_i, k_i \rangle$ *is the initial state of each entity* e_i, $i \leq n$. *Let* $\mathcal{B}_{e_i}$ *be the $\mathcal{S}$BIP transformation and* $\mathbf{X}_{\mathsf{init}}^i$ *be the initial valuation of the entity* e_i *and let* Γ *be the corresponding set of interactions. Lastly, let* $\mathcal{M} = (Q, \pi, P, q_0)$ *be the semantics of* $\langle\ll\rangle(\Gamma(\mathcal{B}_1, \ldots, \mathcal{B}_n))$ *with the initial valuation* $\mathbf{X}_{\mathsf{init}}^1 \sqcup \cdots \mathbf{X}_{\mathsf{init}}^n$. *Then there exists* $\mathcal{R} \subseteq S \times Q$ *a symmetric relation such that*

- $(s_0, q_0) \in \mathcal{R}$;
- *if* $(s, q) \in \mathcal{R}$ *then for all* $s' \in S$ *and* $s \xrightarrow[l]{[n]} s' \in T$ *there exists* $q' \in Q$ *and* $q \xrightarrow{\gamma} q' \in \pi$ *with* $\mathbb{P}(q \xrightarrow{\gamma} q') = n$ *such that* $(s', q') \in \mathcal{R}$.

Lack of space prevents us from writing the proof here, but it is available in the appendix. Moreover, the proof is a straightforward, albeit tedious, application of the definitions. As we noted at the beginning of the section, the correspondence between IoT processes and their $\mathcal{S}$BIP components is kept throughout the transformation thanks to the labeling functions. The stochastic behavior of

$\mathcal{S}$BIP is implemented in IoT by the CHOICE rule. Lastly priorities in $\mathcal{S}$BIP are implemented by the rules ParState_Tau and ParState_Interaction.

As an example, we show in Fig. 6 the transformation of the IoT model of the Smart Hospital in an $\mathcal{S}$BIP component.

6   Evaluating the Probability of an Attack

In this section we use executions of an IoT system to evaluate the probability of an attack. Thanks to Theorem 1, instead of reasoning on an IoT system, we can use the corresponding $\mathcal{S}$BIP system.

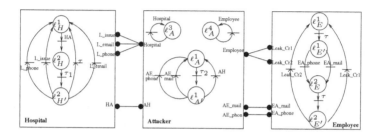

Fig. 6. Transformation of the Smart-Hospital example

We employ two SMC techniques. We first use the Monte Carlo method, which consists of sampling executions and then estimating the probability of an attack, based on the number of executions for which the attack was successful. However, as we will see in the next section, the Monte Carlo method requires a large number of simulations for a correct estimate of an event which occurs with probability 10^{-5}. The experimental framework we used does not scale well for a large number of simulations. We therefore employed a second SMC technique, called *importance splitting* [8]. This technique is tailor-made for *rare* events, that is precisely events that occur rarely in a simulation, and for which Monte Carlo does not scale.

Importance splitting requires the decomposition of an execution leading to an attack into a sequence of elements, called levels and denoted l_i, for $i \leq m$ and for a decomposition in m levels. The first level is reached by all executions, while the last level is reached only if the attack succeeds. The levels are ordered $l_0 < \cdots < l_m$ meaning that level l_i is reached only if the previous levels $l_{j<i}$ have been reached before. We write $\mathbb{P}(\sigma > l_i)$ for the probability that l_i was reached during an execution σ. Then $\mathbb{P}(\sigma > l_i) = \mathbb{P}(\sigma > l_i \mid \sigma > l_{i-1})\mathbb{P}(\sigma > l_{i-1})$. Therefore we can compute the probability of the attack as follows: $\prod_{i=1}^{n} \mathbb{P}(\sigma > l_i \mid \sigma > l_{i-1})$. To infer the levels, importance splitting uses a *score* function defined on executions. Intuitively the closer we get to a successful attack, the higher the score.

Attack trees provide an initial decomposition of the attack, on which the score function is defined. The attack tree is transformed into a $\mathcal{S}$BIP component, called a *monitor*. The leaves of the tree are some of the interactions between the Attacker and the other components in the model. The branches of the tree are internal transitions to the monitor component. In a monitor obtained from an attack tree t we associate a Boolean variable, denoted v_n, for every node (or leaf) n of t. The variable associated to a leaf is set to *true* when the associated event occurred in the monitored execution. The variables of each other node are updated according to their corresponding Boolean expression.

We write $h(t)$ for the height of a tree t and $d(n,t)$ for the depth of node n in t. The score of an execution is computed as $\mathsf{score} = h(t) - d(n,t)$, where n is the highest node for which v_n is *true*.

Definition 11 (Monitor). *The monitor $M_t = (P, V, N)$ of an attack tree t is defined as follows:*

- *$P = \{p_{SR}, p_{LC}, p_{score}\}$ consists of two ports used for observing the SR and LC interactions and of a third port used for an internal transition that updates the score;*
- *$V = V^d \cup V^p$ where $V^d = \{\mathsf{score}\} \cup \{v_n \mid n$ is a node of t$\}$ and $V^p = \emptyset$;*
- *$N = (\{l_0\}, \{l_0\}, T)$ is a Petri-Net with only one place l_0 and with $T = \{(l_0, \langle p, true, f \rangle, l_0) \mid p \in P\}$ where f updates the variables v_n and score.*

For each interaction in an $\mathcal{S}$BIP system, we add a port of the monitor to the interaction. In this manner, the monitor can observe the system and update its Boolean variables accordingly.

Definition 12. *Let Γ be a set of interactions of a $\mathcal{S}$BIP system. The set of interactions Γ' between the $\mathcal{S}$ BIP system and a monitor $M_t = (P, V, N)$ is defined as follows: for all $\gamma = (\{a, a'\}, G_\gamma, F_\gamma) \in \Gamma$, let $(\{a, a', p\}, G_\gamma, F_\gamma \sqcup f) \in \Gamma'$ where p and f are defined as follows:*

- *if $a = e_1 \xrightarrow{c}_{v} e_2$ then $p = p_{SR}$ and $f = \{v_n := true,$ if $v_n \in V\}$, for $n = (SR, v)$;*
- *if $a = e_1 \xrightarrow{\quad}_{v} e_2$ then $p = p_{SR}$ and $f = \{v_n := true,$ if $v_n \in V\}$ for $n = (LC, v)$;*

7 Implementation and Experiments

In this section we describe the tool chain we implemented and some experiments based on the Smart Hospital example. We provide all resources necessary for replicating the experiments at http://iot-modeling.gforge.inria.fr.

In the diagram of Fig. 7, we describe the tool chain we implemented. The user provides an IoT system and an attack tree in the form of a *json* file. We implemented two parsers, "IoT-to-BIP" and "json-to-BIP", that transform the IoT model and the attack tree into two $\mathcal{S}$BIP files. The $\mathcal{S}$BIP files are compiled into a BIP executable. The BIP simulation engine runs the executable and interacts with Plasma [4], the statistical model checker we used.

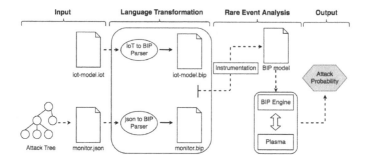

Fig. 7. Tool-set overview

	Model A				Model B			
	Monte Carlo		Importance Splitting		Monte Carlo		Importance Splitting	
Nb of	Result	Time	Result	Time	Result	Time	Result	Time
Simulations	$(\times 10^{-5})$	(s)	$(\times 10^{-5})$	(s)	$(\times 10^{-5})$	(s)	$(\times 10^{-6})$	(s)
1000	0	6	6,7	8	0	6	6,4	8
10000	8,0	15	5,2	30	0	15	11,9	30
100000	6,7	255	7,0	1349	0	251	5,8	1088
1000000	6,4	10410	4,0E-5	26330	7,5	10423	4,3E-6	27013

Fig. 8. Experiments: In model B the *leaks* are twice less probable than in model A.

The results of our experiments are shown in Fig. 8. The model we used is based on our running example, the Smart Hospital.

We used two variants of our IoT model to calculate the probabilities of the success of an attack using the Monte Carlo and the importance splitting methods. The results and times presented in Fig. 8 are obtained by averaging outcomes of 10 iterations. We observe that importance splitting gives a correct estimate from 1000 simulations, whereas the Monte Carlo method needs 100 times (1000 times for Model B) more simulations. However the importance splitting methods does not behave well when running on a large number of simulations. Therefore we argue that both methods are useful and complement each other in our analysis: Monte Carlo for estimating a probability n when we can produce around $10/n$ simulations, and importance splitting for experiments with fewer simulations.

8   Related Work

In this section we compare our work with some other works in the literature. First we explain why we defined our own formal language for IoT. Then we look at other works that used attack trees and model checking to analyse security properties of a system.

The formal language we propose for specifying IoT systems is an extension of [2] to probabilistic systems, inspired by the probabilistic CCS [17]. However our treatment of probabilities is slightly different (see rule CHOICE in Fig. 2).

An IoT model in our language is transformed into a $\mathcal{S}$BIP system, which has a given semantics. The operational semantics we defined therefore reflects the way probabilities are handled in $\mathcal{S}$BIP. Moreover, there are also some difference in terms of expressiveness between CCS and our language. We designed our language such that entities can store data (and thus have some notion of state) and we distinguish between two types of interactions: either safe ones or interactions that contain leaks of sensitive data. We motivate these features of our language in Sect. 2.

Attack trees are used extensively for modeling security attacks, including internet attacks as in [16]. In this work we use attack trees in combination with a model checker to monitor a system's execution and infer the probability of a successful attack. We therefore restrict our survey to related works that do a similar analysis.

In [6,10] attack trees are translated into timed and stochastic automata. Model checking is then used to infer various properties of an attack, such as its probability of success or its cost. In [7] attack trees are translated into stochastic Petri Nets such that attacks can then be simulated. Compared to these works, we model both the attacks, using the attack trees, *and* the system on which the attacks are carried. In our approach it is the model of the IoT system that is decorated with probabilities and is simulated, and not the attack trees. Moreover, the tools and the approach we propose can also tackle rare events in security issues.

The authors of [13] presented a formal model to describe IoT systems and ensure its functional correctness. Their proposed approach is similar to ours, however the language we propose is more general and, similar to above, we allow for a stochastic analysis that handles rare attacks.

In [14] the *importance sampling* technique for rare events is applied to *fault-trees*, a variant of attack trees. The technique consists of computing a new distribution that makes the rare event more frequent. The results are then adjusted w.r.t. the difference between the normal distribution and the importance sampling one. Our method based on importance splitting, is better suited for attack trees, as the intermediate steps leading to a rare event correspond to the nodes in the tree leading to a successful attack. Moreover, there are no additional steps as in the case of importance sampling.

Formal semantics for attack trees have been proposed in [9,12] and we draw inspiration from them when introducing attack trees in Sect. 3.

9 Conclusion

In this paper, we proposed a sound probabilistic framework for modeling IoT systems and verifying its security using attack trees. The approach consists on transforming a high level IoT model and its attack tree into a $\mathcal{S}$BIP model. We showed on a complex example how to estimate the probability of success of an attack using SMC techniques.

Acknowledgements. We would like to thank Axel Legay for his helpfull suggestions on importance splitting, and Jean Quilbeuf for his technical help in the tool implementation.

Appendix

Counting Functions for the Operational Semantics of IoT

Definition 13 (Counting τ transitions from a state). *The functions* $\mathsf{count}_\tau : State \to \mathbb{N}$ *and* $\mathsf{count\_proc}_\tau : Proc \to \mathbb{N}$ *are defined as follows:*

$$\mathsf{count}_\tau(s|t) = \mathsf{count}_\tau(s) + \mathsf{count}_\tau(t)$$
$$\mathsf{count}_\tau(\langle P, k \rangle) = \mathsf{count\_proc}_\tau(P)$$
$$\mathsf{count\_proc}_\tau(0) = 0$$
$$\mathsf{count\_proc}_\tau(\alpha.P) = 1 \; if \; \alpha \neq \tau$$
$$0 \; if \; \alpha = \tau$$
$$\mathsf{count\_proc}_\tau(\sum \alpha_i.P_i) = 1$$
$$\mathsf{count\_proc}_\tau(P \mid Q) = \mathsf{count\_proc}_\tau(P) + \mathsf{count\_proc}_\tau(Q).$$

For counting the number of interactions, we have first to rewrite a state into a *canonical* form:

$$s \equiv s_S \mid s_R \mid s_L \mid s_C \qquad where \qquad \begin{aligned} s_S &= \langle P_1^S, k_1^S \rangle \mid \cdots \langle P_{nS}^S, k_{nS}^S \rangle \\ s_R &= \langle P_1^R, k_1^R \rangle \mid \cdots \langle P_{nR}^R, k_{nR}^R \rangle \\ s_L &= \langle P_1^L, k_1^L \rangle \mid \cdots \langle P_{nL}^L, k_{nL}^L \rangle \\ s_C &= \langle P_1^C, k_1^C \rangle \mid \cdots \langle P_{nC}^C, k_{nC}^R \rangle \end{aligned}$$

and where $P_i^S \equiv a.P$ and the action a is a send; nS is the number of processes of the form above in s. Similarly we define the rest of the processes. Note that if we cannot rewrite a state in this form then the rule PARSTATE_INTERACTION cannot be applied (any internal or sum transitions have priority over the interactions). Moreover entities can only communicate with other entities, that is interactions are not defined internally to an entity. We therefore only need to count interactions between entities.

The function $\mathsf{count}_{SR,LC}$ uses an auxiliary function $\bar{\cdot}$: action $\to$ action which defines an action $\bar{a}$ which can synchronise with a using the rules SENDRECEIVE or LEAKCOLLECT.

Definition 14. *Let $s \equiv s_S \mid s_R \mid s_L \mid s_C$ be a state in a canonical form. The function* $\text{count}_{SR,LC} : State \to \mathbb{N}$ *is defined on s as follows:*

$$\text{count}_{SR,LC}(s_S \mid s_R \mid s_L \mid s_C) = \text{count}_{SR}(s_S, s_R) + \text{count}_{LC}(s_L \mid s_C)$$
$$\text{count}_{SR}(\langle a.P, k \rangle \mid s, t) = \text{count}(a, t) + \text{count}_{SR}(s, t)$$
$$\text{count}_{LC}(\langle a.P, k \rangle \mid s, t) = \text{count}(a, t) + \text{count}_{LC}(s, t)$$
$$\text{count}(a, \langle b.P, k \rangle \mid t) = 1 + \text{count}(a, t) \ \text{if } a = \overline{b}$$
$$= \text{count}(a, t) \ \text{otherwise}$$

Proof of Theorem 1

Lemma 1. *Any two congruent IoT states have the same transformation in $\mathcal{S}BIP$ systems.*

Proof. We proceed by cases on the congruence relation. First consider the congruence relation on states: For the monoid laws on $\mid$, note that the transformation results in a set of atomic components and therefore the order of states in the parallel composition does not matter. In the case where processes are congruent, we distinguish two subcases: (i) Threads in a parallel composition translate into tuples of states in the transformation of a process (Definition 9) where the order of the states does not matter; (ii) For the rest we use the fact that inside an atomic component the states that have congruent labels are identified.

Proof (Theorem 1). Let $e_1, \cdots, e_n$ be n entities of an IoT system (S, L, T, s_0) with the initial states $\langle P_1, k_1 \rangle, \cdots \langle P_n, k_n \rangle$. $\mathcal{B}_{e_i} = (P_i, V_i, N_i)$ with $N_i = (L_i, L_{i,0}, T_i, F_i)$, is the transformation of the current state of the entity e_i, for $i \leq n$. Also let $V_i = V_i^p \cup V_i^d$. We write (P, Q, π, q_0) for the semantics of $\langle \ll \rangle \Gamma(\mathcal{B}_{e_1}, \cdots \mathcal{B}_{e_n}) = (\Gamma, \mathbf{V}, \mathbf{N})$ with $\mathbf{N} = (\mathbf{L}, \mathbf{L_0}, \mathbf{T}, \mathbf{F})$. Lastly $\mathbf{X_{init}}$ is the initial valuation.

To construct the relation $\mathcal{R} \subseteq S \times Q$ required by the theorem, we first set some notations and constraints below. Informally, these constraints establish the relation between the processes and knowledge functions in states of S and the markings and the valuations, respectively, in states of Q.

1. **Correspondence between Processes and Markings.** For a thread T let us write m_T for the marking associated with T and defined as follows:

   $$m_T(l) = 1 \ \text{if } \ell(l) = T \text{ or } \ell(l) = U^*, U = [n]T + T', \quad \text{for some threads } T', U$$
   $$0 \text{ otherwise}$$

   where (P, V, N) and $N = (\mathcal{L}, \mathcal{L}_0, \mathcal{T}, \mathcal{F})$ is obtained as in Definition 8 and where $l \in \mathcal{L}$. For a process $P = T_1 \mid \cdots \mid T_m$, let us write m_P for the marking associated with P and defined as $m_{T_1} + \cdots + m_{T_m}$.

2. **Correspondence between Knowledge and the Deterministic Variables.** From Definition 8 it follows that for each thread T_j in a process P_i we define the set $V_i^d = \{v_c \mid c \text{ is a protocol used in } T_j\}$. From Definition 9 then

the set of variables of $P_i = T_1 \mid \cdots \mid T_m$ is $\cup_{j \leq m} V_j = \{v_c \mid c$ is a protocol used in $P_i\}$.

Then, if $\mathbf{X_i}$ the current valuation of entity e_i, we require that $\mathbf{X_i}(v_c) = k_i(c)$, for $i \leq n$, $c \in C$ and $v_c \in V_i^d$. Recall that we write C for the set of protocols used in the IoT system and k_i for the knowledge function of an entity e_i.

3. **Correspondence between Probabilistic Choices in Processes and the Random Variables.** For every summation thread U in a process P_i, we have that there exists a random variable $v_U \in V_i^p$, by Definition 8. Moreover, if T a thread of P_i, belongs to a summation, i.e. $U = [n]T + T'$, for some threads T', U, then for the current valuation $\mathbf{X_i}$ we have that $\mathbf{X_i}(v_U) = T$. For a process $P = T_1 \mid \cdots \mid T_m$ we use Definition 9 and have that V^p is the disjoint union of all V_j^p, where V_j^p is the set of random variables for T_j, $j \leq m$.

We define the following relation between the states of S and the states of Q:

$$\mathcal{R} = \Big\{ \big(\langle P_1, k_1 \rangle \mid \cdots \mid \langle P_n, k_n \rangle, (m = m_{P_1} + \cdots m_{P_n}, \mathbf{X} = \mathbf{X_1} \sqcup \cdots \sqcup \mathbf{X_n}) \big) \mid$$
$$\text{the conditions } 1-3 \text{ above hold} \Big\}.$$

We show that $\mathcal{R}$ is the relation required in Theorem 1. First we have to show that $(s_0, q_0) \in \mathcal{R}$.

We use Definition 8 from which we have that $\mathcal{L}_0 = \{l_T\}$ is the initial place in the transformation of a thread T. Then, by Definition 9, $L_0 = \uplus_{j \leq m} \mathcal{L}_0^j = \uplus_{j \leq m} \{l_{T_j}\}$ is the initial set of places in the transformation of a process $P = T_1 \mid \cdots \mid T_m$. From Definition 7 it follows that $\mathbf{L_0} = \uplus_{i \leq n} L_{0,i}$. By Definition 5 the initial marking in $q_0 = (m_0, \mathbf{X_{init}})$ is defined as $m_0(l) = 1 \iff l \in \mathbf{L_0}$ and 0 otherwise. Hence we can write $m_0 = m_{P_1} + \cdots + m_{P_n}$. This shows condition 1 of $\mathcal{R}$.

From Definition 9 we have that for each entity e_i, $\mathbf{X_{init}}(v_c) = k_i(c)$, for all protocols c used by e_i. From Definition 7 the set of variables of the composed $\mathcal{B}_{e_i}$ components is the disjoint union V_i, i.e. $\mathbf{V} = \uplus_{i \leq n} V_i$, in particular $\mathbf{V}^d = \uplus_{i \leq n} V_i^d$. Then a valuation for $\mathbf{V}$ is the disjoint composition of the individual valuations for V_i, from which it follows the required decomposition of $\mathbf{X_{init}}$ in $q_0 = (m_0, \mathbf{X_{init}})$. Therefore condition 2 of $\mathcal{R}$ holds. For condition 3 to hold suffices to note that there is no probabilistic choice made yet in any process and therefore there is no correspondence to show. We can take any initial valuation we want for the random variables.

Let us now suppose that $(s, q) \in \mathcal{R}$ and that $s \xrightarrow[l]{[n]} s'$, for some label $l \in L$, some probability n and state $q' \in Q$. We have to show that there exists $q' \in Q$ and $q \xrightarrow{p} q' \in \pi$ with $\mathbb{P}(q \xrightarrow{p} q') = n$ such that $(s', q') \in \mathcal{R}$. We reason by cases on the label l of the transition $s \xrightarrow[l]{[n]} s'$.

– Let $l = SR : v$ or $l = LC : v$; then let e_1 and e_2 be the two communicating entities. Using Lemma 1 we can rewrite the transition as follows:

$$s = \langle P_1, k_1 \rangle \mid \langle P_2, k_2 \rangle \mid \langle P_3, k_3 \rangle \mid \ldots \mid \langle P_n, k_n \rangle \xrightarrow[l]{[1/m]}$$

$$s' = \langle Q_1, k_1' \rangle \mid \langle Q_2, k_2' \rangle \mid \langle P_3, k_3 \rangle \mid \ldots \mid \langle P_n, k_n \rangle$$

where we can decompose $P_1 \equiv_P a_1.T_1 \mid P_1'$ and $P_2 \equiv_P a_2.T_2 \mid P_2'$, $Q_1 \equiv_P T_1 \mid P_1'$ and $Q_2 \equiv_P T_2 \mid P_2'$, again by Lemma 1 and from the rules of Fig. 2. Here we suppose w.l.o.g. that a_1 and a_2 are the two synchronizing actions in P_1 and P_2, respectively. Also suppose w.l.o.g. that a_1 is a send (or a leak) and that a_2 is a receive (or a collect). Let c be the protocol used for the communication in case $l = SR : v$.

From $(s, q) \in \mathcal{R}$ we have that $q = (m_{P_1} + \cdots m_{P_n}, \mathbf{X_1} \sqcup \cdots \sqcup \mathbf{X_n})$ and that $m_{P_i} = m_{a_i.T_i} + m_{P_i'}$, for $i \leq 2$. Also from condition 1 of $\mathcal{R}$, $m_{a_i.T_i} = \{l_i\}$ with either $\ell(l_i) = a_i.T_i$, or $\ell(l_i) = U_i^*$, for some summation threads U_1, U_2.

- If $\ell(l_1) = a_1.T_1$ then we use the transformation of Definition 8 to show that there exists the place $l_1' \in L_1$, with $\ell(l_1') = T_1$ and the transition $t_1 = (\{l_{a_1.T_1}\}, \langle a_1, g_1 = true, f_1 \rangle, \{l_{T_1}\})$ in $\mathcal{B}_1$.
  * If $\ell(l_2) = a_2.T_2$ then as above, there exists $l_2' \in L_2$, with $\ell(l_2') = T_2$ and the transition $t_2 = (\{l_{a_2.T_2}\}, \langle a_2, g_2 = true, f_2 \rangle, \{l_{T_2}\})$ in $\mathcal{B}_2$.
  * $\ell(l_2) = U_2^*$, with $U_2 = [n_2]a_2.T_2 + U_2'$, for some threads U_2, U_2'. As in the case above, from Definition 8 we have that there exists the places $l_2' \in L_2$ with $\ell(l_2') = T_2$. We also have, from condition 3 of $\mathcal{R}$ that there exists a random variable $v_{U_2} \in V_2^p$ with $\mathbf{X}(v_{U_2}) = a_2.T_2$. Moreover we have the transition $t_2 = (\{l_{U_2^*}\}, \langle a_2, g_2 = (v_{U_2} == a_2.T_2), f_2 \rangle, \{l_{T_2}\})$ in $\mathcal{B}_2$.
- the other case is similar.

Note that in all cases above, $f_i = \{v := v \mid v \in V^d\}$ with $R_i^p = \emptyset$, $i \leq n$. Using Definition 10 we have that there exists an interaction $\gamma = (\{a_1, a_2\}, G, F)$ such that
- If $l = SR : v$ then $G = (\exists x \in v_c^1$ such that $x \in v_c^2)$ for $v_c^1 \in V_1$ and $v_c^2 \in V_2$.
- If $l = LC : v$ then $G = true$.

Also, $F = \{v_{c'}^2 := v_{c'}^2 \cup \{v'\} \mid \mathsf{protocol}(v') = c', v_{c'}^2 \in V_2\}$ for both $l = SR : v$ and $l = LC : v$.

We now use Definition 7 and have that there exists the transition

$$\underline{T} = (\{l_1, l_2\}, \langle \gamma, g_1 \wedge g_2 \wedge G, (f_1 \sqcup f_2) \circ F \rangle, \{l_1', l_2'\}) \in \mathbf{T}.$$

We have to show that the guard $g = g_1 \wedge g_2 \wedge G$ holds for the current valuation $\mathbf{X}$:
- If $g_1 = (v_{U_1} == a_1.T_1)$ then $\mathbf{X}(g_1)$ holds from condition 3 of $\mathcal{R}$; otherwise $g_1 = true$. We proceed similarly for g_2.
- If $l = SR : v$ then $G = (\exists x \in v_c^1$ such that $x \in v_c^2)$ for $v_c^1 \in V_1$ and $v_c^2 \in V_2$. From condition 2 of $\mathcal{R}$ we have that $\mathbf{X}(v_c^i) = k_i(c)$, $i \leq n$. Then the guard holds as it is the condition of rule SENDRECEIVE in Fig. 2. If $l = LC : v$ then $G = true$.

The transitions above are allowed to proceed by the priority order $\ll$ (see text after Definition 7) only if there is no internal transition available. This is the case as ensured by the rule PARSTATE_INTERACTION in Fig. 2. Therefore, by Definition 5, there exists the transition

$$q = (m_{P_1} + m_{P_2} + \cdots m_{P_n}, \mathbf{X_1} \sqcup \cdots \sqcup \mathbf{X_n}) \xrightarrow{\gamma} q' = (m', \mathbf{X'})$$

where we have to show that conditions 1-3 of $\mathcal{R}$ hold. For condition 1 we have to show that $m' = m_{Q_1} + m_{Q_2} + \cdots m_{P_n}$. Using Definition 5 it follows that

$$m' = m - {}^\bullet\underline{T} + \underline{T}^\bullet = m - \{l_1, l_2\} + \{l'_1, l'_2\}.$$

As $\mathbf{L_0} = \uplus_{i \leq n} L_{0,i}$, from Definition 7, it follows that

$$m' = (m_{P_1} - \{l_1\} + \{l'_1\}) + (m_{P_2} - \{l_2\} + \{l'_2\}) + \cdots + m_{P_n}.$$

Using condition 1 of $\mathcal{R}$ on m_{P_1} and m_{P_2} we have that $m_{P_1} - \{l_1\} + \{l'_1\} = m_{Q_1}$ and similarly for m_{Q_2}.

Let us now show condition 2, i.e. $\mathbf{X'} = \mathbf{X'_1} \sqcup \mathbf{X'_2} \sqcup \cdots \sqcup \mathbf{X_n}$ and $\mathbf{X'_i}(v_{c'}) = k'_i(c')$, $i \leq 2$. Using the function F above we have that $\mathbf{X'_i}(v_{c'}) = \mathbf{X_i}(v_{c'}) \cup \{v\}$. From rules SENDRECEIVE and LEAKCOLLECT we also get that $k'_i(c') = k_i(c) \cup \{v\}$, $i \leq 2$.

As $R_1^p = R_2^p = \emptyset$ condition 3 is trivial.

Lastly, the two transitions have the same probability: $|\mathsf{Enabled}(m; \mathbf{X})| = m$ by Lemma 1, and therefore $\mathbb{P}(q \xrightarrow{p} q') = 1/m$.

- Let $l = \tau$; let e_1 be the entity that triggers the internal transition. Using Lemma 1 we can rewrite the states in the transition as follows:

$$s = \langle P_1, k_1 \rangle \mid \langle P_2, k_2 \rangle \mid \langle P_3, k_3 \rangle \mid \ldots \mid \langle P_n, k_n \rangle \xrightarrow[l]{[n]}$$

$$s' = \langle Q_1, k'_1 \rangle \mid \langle P_2, k_2 \rangle \mid \langle P_3, k_3 \rangle \mid \ldots \mid \langle P_n, k_n \rangle.$$

There are two possibilities: either $P_1 \equiv_P \sum_{i \in I_1} a_i.T_i \mid P'_1$ where $Q_1 = a_1.T_1$ w.l.o.g. or $P_1 \equiv_P \tau.T_1 \mid P'_1$ with $Q_1 = T_1$. We write $U = \sum_{i \in I_1} a_i.T_i$ or $U = \tau.T_1$ depending on which of the two cases we are.

From $(s, q) \in \mathcal{R}$ we have that $q = (m_{P_1} + \cdots m_{P_n}, \mathbf{X_1} \sqcup \cdots \sqcup \mathbf{X_n})$ and that $m_{P_1} = \{l\} + m_{P'_1}$, $\ell(l) = U$. We use the transformation of Definition 8 to show that there exists the place $l' \in L_1$ and the transition $t = (\{l\}, \langle \tau, g = true, f \rangle, \{l'\})$ in $\mathcal{B}_1$.

- If $U = \sum_{i \in I_1} a_i.T_i$ then $\ell(l') = U^\star$, $f = \{v := v \mid v \in V_1^d\}$ and $R^p = \{v_U\}$.
- If $U = \tau.T_1$ then $\ell(l') = T_1$, $f = \{v := v \mid v \in V_1^d\}$ and $R^p = \emptyset$.

Using Definition 10 we have that there exists an interaction $\gamma = (\{\tau\}, G = true, F)$ with $F = \{v := v \mid v \in V_1^d\}$.

From Definition 7 there exists the transition

$$\underline{T} = (\{l\}, \langle \gamma, g_1 \wedge G = true, f \circ F \rangle, \{l'\}) \in \mathbf{T}.$$

The guard trivially holds and we obtain the transition

$$q = (m_{P_1} + + \cdots m_{P_n}, \mathbf{X_1} \sqcup \cdots \sqcup \mathbf{X_n}) \xrightarrow{\gamma} q' = (m', \mathbf{X}')$$

where we have to show that conditions 1-3 of $\mathcal{R}$ hold. As in the first case, condition 1 follows from $m' = m - \{l\} + \{l'\} = m_{Q_1} + \cdots m_{P_n}$. Condition 2 trivially hold as the update functions f and F are the identity and therefore $\mathbf{X_1}' = \mathbf{X_1}$. Indeed the knowledge function of k_1 is not modified by the rules CHOICE or INTERNAL.

To show condition 3 we use Definition 8 from which we have that there exists $v_U \in V_1^p$, $v_U \sim \mu$, where $\mu(a_1.T_1) = n_1$. Then we can take $\mathbf{X}'(v_U) = a_1.T_1$. We also this argument to show that the two transitions have the same probabilities: by Lemma 1, $|\mathsf{Enabled}(m; \mathbf{X})| = m$ and therefore $\mathbb{P}(q \xrightarrow{p} q') = 1/m \times n_1$.

Hereafter we prove the similarity of the IoT system to its corresponding $\mathcal{S}$BIP model. Let us suppose that $(q, s) \in \mathcal{R}$ and that $q \xrightarrow{\gamma} q'$, for a transition labelled with γ, where $q, q' \in Q$. We have to show that there is a state $s' \in S$ with $s \xrightarrow[l]{[n]} s'$, for some label $l \in L$, such that $(s', q') \in \mathcal{R}$. We define $s = \langle P_1, k_1 \rangle \mid \langle P_2, k_2 \rangle \mid \cdots \mid \langle P_n, k_n \rangle$. Here we also reason by cases: whether the transition is an interaction between two components $\mathcal{B}_{e_1}$ and $\mathcal{B}_{e_2}$ or an internal transition.

- We consider the communication is an interaction $\gamma = (\{a_1, a_2\}, G, F)$ between $\mathcal{B}_{e_1}$ and $\mathcal{B}_{e_2}$:

$$q = (m_{P_1} + m_{P_2} + \cdots m_{P_n}, \mathbf{X_1} \sqcup \mathbf{X_2} \sqcup \cdots \sqcup \mathbf{X_n}) \xrightarrow{\gamma} q' = (m', \mathbf{X}')$$

As it is an interaction between two entities, from Definition 7 we have that there exists the transitions $t_i = (m_i, \langle p_i, g_i, f_i \rangle, m_i') \in T_i$, for $i \in \{1, 2\}$. From the Definition 10, $m_i = m_{P_i}$, $p_i = a_i$, $g_i = true$ and f_i are the constant update functions. From $(q, s) \in \mathcal{R}$ we have that $m_{P_1} = m_{a_1.T_1} + m_{P_1'}$, $m_{P_2} = m_{a_2.T_2} + m_{P_2'}$ with $P_1 = a_1.T_1 \mid P_1'$ and $P_2 = a_2.T_2 \mid P_2'$. Moreover, from the Definition 5 there exists the transition

$$\underline{T} = (m_{P_1} + m_{P_2}, \langle \{a_1, a_2\}, g_1 \wedge g_2 \wedge G, (f_1 \sqcup f_2) \circ F \rangle, m_{Q_1} + m_{Q_2}) \in \mathbf{T}$$

with $m_{Q_1} = m_{T_1} + m_{P_1'}$, $m_{Q_2} = m_{T_2} + m_{P_2'}$.

We distinguish between the two types of interactions:

- $a_1 = e_1 \xrightarrow[v']{c} e_2$ and there exists $a_2 \in \mathsf{Actions}$ such that $a_2 = e_2 \xleftarrow{c} e_1$,
- or $a_1 = e_1 \xrightarrow[v']{} e_2$ and there exists $a_2 \in \mathsf{Actions}$ such that $a_2 = e_2 \leftarrow e_1$

Following the Definition 10 we have the following guards:

- if $G = (\exists x \in v_c^1$ such that $x \in v_c^2)$ for $v_c^1 \in V_1$ and $v_c^2 \in V_2$ then $l = SR$
- if $G = true$ then $l = LC$

We can then apply the rules SENDRECEIVE or LEAKCOLLECT from Fig. 2. Hence we derive an interaction between e_1 and e_2 exists for which we have to show that conditions 1–3 of $\mathcal{R}$ hols.

$$s = \langle P_1, k_1 \rangle \mid \langle P_2, k_2 \rangle \mid \ldots \mid \langle P_n, k_n \rangle \xrightarrow[l]{[n]}$$
$$s' = \langle Q_1, k_1' \rangle \mid \langle Q_2, k_2' \rangle \mid \ldots \mid \langle P_n, k_n \rangle.$$

From above, it follows that $m' = m_{Q_1} + m_{Q_2} + \cdots m_{P_n}$, which is the first condition of $\mathcal{R}$.

In the interaction γ, we apply the update function $F = \{v_{c'}^2 := v_{c'}^2 \cup \{v'\} \mid \text{protocol}(v') = c', v_{c'}^2 \in V_2\}$ for both $l = SR : v$ and $l = LC : v$, then $\mathbf{X_i'}(v_{c'}) = \mathbf{X_i}(v_c) \cup \{v\}$. Therefore we can write $\mathbf{X}' = \mathbf{X_1'} \sqcup \mathbf{X_2'} \cdots \mathbf{X_n}$. With the interaction $s \xrightarrow[l]{[n]} s'$, we apply rules SENDRECEIVE or LEAKCOLLECT from Fig. 2 where $k_i'(c') = k_i(c) \cup \{v\}$. Hence the condition 2 hols, i.e. $\mathbf{X_i'}(v_{c'}) = k_i'(c')$. With the execution of the γ interaction, the probabilistic distribution $R_1^p = R_2^p = \emptyset$, and from the SENDRECEIVE or LEAKCOLLECT from Fig. 2 is the same, then the condition 3 trivially hols. The two transitions have the same probability: $\mathbb{P}(q \xrightarrow{p} q') = 1/m$ by Lemma 1, and therefore $|\text{Enabled}(m; \mathbf{X})| = m$.

– We consider the transition to be an internal transition τ in component $\mathcal{B}_{e_1}$. From Lemma 1 we can write the transition:

$$q = (m_{P_1} + m_{P_2} + \cdots m_{P_n}, \mathbf{X_1} \sqcup \mathbf{X_2} \sqcup \cdots \sqcup \mathbf{X_n}) \xrightarrow{\gamma} q' = (m', \mathbf{X}')$$

where $s = \langle P_1, k_1 \rangle \mid \langle P_2, k_2 \rangle \mid \cdots \mid \langle P_n, k_n \rangle$, from $(q, s) \in \mathcal{R}$, we distinguish two cases of transition execution:

- A probabilistic choice: $m_{P_1} = \{l\} + m_{P_1'}$ where $\ell(l) = \sum_{i \in I} [n_i] a_i.T_i$ and $P_1 = \sum_{i \in I} a_i.T_i \mid P_1'$. From the transformation of Definition 8, the transition

$$t = (\{l_T\}, \langle \tau, \text{true}, f^\star \rangle, \{l_{T^\star}\}) \in T_1$$

can be executed where $f^\star = (\{v := v \mid v \in V^d\}$ and $R^p = \{v_T\})$. From relations of Fig. 2, there exists a CHOICE transition in IoT system such that

$$s = \langle P_1, k_1 \rangle \mid \langle P_2, k_2 \rangle \mid \ldots \mid \langle P_n, k_n \rangle \xrightarrow[l]{[n_1]}$$
$$s' = \langle Q_1, k_1' \rangle \mid \langle P_2, k_2 \rangle \mid \ldots \mid \langle P_n, k_n \rangle$$

where $Q_1 = a_1.T_1$. Now we can verify if the conditions 1–3 of R holds. We have that $m_{Q_1} = \{\ell\}^\star$ and $m' = m_{Q_1} + m_{P_2} + \cdots m_{P_n}$. As the update function f is the identity function the condition 2 trivially hold and the knowledge $k_1' = k_1$. To show condition 3, we note that there exists $v_{T_1} \in V_1^p$, $v_{T_1} \sim \mu$ such that $\mathbf{X}'(v_{T_1}) = a_1.T_1$. We use Definition 8 from which we have that where $\mu(a_1.T_1) = n_1$.

- An internal transition: $m_{P_1} = m_{\tau.T_1} + m_{P_1'}$ and $P_1 = \tau.T_1|P_1'$. From the transformation of Definition 8, the transition $\underline{T} = (\{l_{a.T}\}, \langle a, \text{true}, f \rangle, \{l_T\})$ can be executed where $f = (\{v := v \mid v \in V^d\}$ and $R^p = \emptyset)$. From relations of Fig. 2, there exists an INTERNAL transition in IoT system such that

$$s = \langle P_1, k_1 \rangle \mid \langle P_2, k_2 \rangle \mid \ldots \mid \langle P_n, k_n \rangle \xrightarrow[l]{[n]}$$
$$s' = \langle Q_1, k_1' \rangle \mid \langle P_2, k_2 \rangle \mid \ldots \mid \langle P_n, k_n \rangle$$

where $Q_1 = T_1|P_1'$. Now we can verify if the conditions 1–3 of R holds. We have that $m_{Q_1} = m_{T_1} + m_{P_1'}$ and $m' = m_{Q_1} + m_{P_2} + \cdots m_{P_n}$.

As the update function f is the identity function the condition 2 trivially hold and the knowledge $k_1' = k_1$. Then $\mathbf{X'} = \mathbf{X_1'} \sqcup \mathbf{X_2} \sqcup \cdots \sqcup \mathbf{X_n}$. Likewise, since $R^p = \emptyset$ the condition 3 trivially holds.

References

1. Basu, A., Bozga, M., Sifakis, J.: Modeling heterogeneous real-time components in BIP. In: SEFM (2006). https://doi.org/10.1109/SEFM.2006.27
2. Beaulaton, D., et al.: A language for analyzing security of IoT systems. In: SoSE (2018). https://doi.org/10.1109/SYSOSE.2018.8428704
3. Bensalem, S., Bozga, M., Delahaye, B., Jegourel, C., Legay, A., Nouri, A.: Statistical model checking QoS properties of systems with SBIP. In: Margaria, T., Steffen, B. (eds.) ISoLA 2012. LNCS, vol. 7609, pp. 327–341. Springer, Heidelberg (2012). https://doi.org/10.1007/978-3-642-34026-0_25
4. Boyer, B., Corre, K., Legay, A., Sedwards, S.: PLASMA-lab: a flexible, distributable statistical model checking library. In: Joshi, K., Siegle, M., Stoelinga, M., D'Argenio, P.R. (eds.) QEST 2013. LNCS, vol. 8054, pp. 160–164. Springer, Heidelberg (2013). https://doi.org/10.1007/978-3-642-40196-1_12
5. ENISA: Smart hospitals, security and resilience for smart health service and infrastructures. Technical report, ENISA (2016)
6. Gadyatskaya, O., Hansen, R.R., Larsen, K.G., Legay, A., Olesen, M.C., Poulsen, D.B.: Modelling attack-defense trees using timed automata. In: Fränzle, M., Markey, N. (eds.) FORMATS 2016. LNCS, vol. 9884, pp. 35–50. Springer, Cham (2016). https://doi.org/10.1007/978-3-319-44878-7_3
7. Dalton, G.C., Mills, R.F., Colombi, J.M., Raines, R.A.: Analyzing attack trees using generalized stochastic Petri nets. In: 2006 IEEE Information Assurance Workshop (2006). https://doi.org/10.1109/IAW.2006.1652085
8. Jegourel, C., Legay, A., Sedwards, S.: Importance splitting for statistical model checking rare properties. In: Sharygina, N., Veith, H. (eds.) CAV 2013. LNCS, vol. 8044, pp. 576–591. Springer, Heidelberg (2013). https://doi.org/10.1007/978-3-642-39799-8_38
9. Kordy, B., Pouly, M., Schweitzer, P.: Computational aspects of attack–defense trees. In: Bouvry, P., Kłopotek, M.A., Leprévost, F., Marciniak, M., Mykowiecka, A., Rybiński, H. (eds.) SIIS 2011. LNCS, vol. 7053, pp. 103–116. Springer, Heidelberg (2012). https://doi.org/10.1007/978-3-642-25261-7_8

10. Kumar, R., et al.: Effective analysis of attack trees: a model-driven approach. In: Russo, A., Schürr, A. (eds.) FASE 2018. LNCS, vol. 10802, pp. 56–73. Springer, Cham (2018). https://doi.org/10.1007/978-3-319-89363-1_4
11. TrapX LAbs: Anatomy of an attack, medjack (medical device attack). Technical report, TrapX Security Inc. (2015)
12. Mauw, S., Oostdijk, M.: Foundations of attack trees. In: Won, D.H., Kim, S. (eds.) ICISC 2005. LNCS, vol. 3935, pp. 186–198. Springer, Heidelberg (2006). https://doi.org/10.1007/11734727_17
13. Ouchani, S.: Ensuring the functional correctness of IoT through formal modeling and verification. In: Abdelwahed, E.H., Bellatreche, L., Golfarelli, M., Méry, D., Ordonez, C. (eds.) MEDI 2018. LNCS, vol. 11163, pp. 401–417. Springer, Cham (2018). https://doi.org/10.1007/978-3-030-00856-7_27
14. Ruijters, E., Reijsbergen, D., de Boer, P.-T., Stoelinga, M.: Rare event simulation for dynamic fault trees. In: Tonetta, S., Schoitsch, E., Bitsch, F. (eds.) SAFECOMP 2017. LNCS, vol. 10488, pp. 20–35. Springer, Cham (2017). https://doi.org/10.1007/978-3-319-66266-4_2
15. Schneier, B.: Secrets & Lies: Digital Security in a Networked World. Wiley, Hoboken (2000)
16. Tidwell, T., Larson, R., Fitch, K., Hale, J.: Modeling internet attacks. In: Proceedings of the 2001 IEEE Workshop on Information Assurance and Security, IA (2001)
17. Vanglabbeek, R., Smolka, S., Steffen, B.: Reactive, generative, and stratified models of probabilistic processes. Inf. Comput. **121** (1995). https://doi.org/10.1006/inco.1995.1123

Attack–Defense Trees for Abusing Optical Power Meters: A Case Study and the OSEAD Tool Experience Report

Barbara Fila$^{(\boxtimes)}$ and Wojciech Wideł

Univ Rennes, INSA Rennes, CNRS, IRISA, Rennes, France
{barbara.fila,wojciech.widel}@irisa.fr

Abstract. Tampering with their power meter might be tempting to many people. Appropriately configured home-placed meter will record lower than the actual electricity consumption, resulting in substantial savings for the household. Organizations such as national departments of energy have thus been interested in analyzing the feasibility of illegal activities of this type. Indeed, since nearly every apartment is equipped with a power meter, the negative financial impact of tampering implemented at a large scale might be disastrous for electricity providers.

In this work, we report on a detailed analysis of the power meter tampering scenario using attack–defense trees. We take various quantitative aspects into account, in order to identify optimal strategies for customers trying to lower their electricity bills, and for electricity providers aiming at securing their infrastructures from thefts. This case study allowed us to validate some advanced methods for quantitative analysis of attack–defense trees as well as evaluate the OSEAD tool that we have developed to support and automate the underlying computations.

1 Introduction

Electricity theft is a widespread practice [19,26] that generates huge financial losses yearly across the world [15,20,28,39], with more than the third of the losses affecting the BRIC countries (Brazil, Russia, India and China) [28]. One of the ways in which electricity is being stolen, is by tampering with power meter in a way that results in the household's or facility's power consumption being underreported. Modern smart meters make identifying crude power meter tampering attempts easier, but remain vulnerable to (not necessarily sophisticated) hacking attacks [32].

This study is concerned with the issue of tampering with power meters. We consider a malicious user whose aim is to reconfigure their power meter, in order to lower the recorded electricity consumption of their household. We extend the *attack tree*-based model of possible behavior of such a user, analyzed by the U.S. Department of Energy in [33], to take possible countermeasures into account. We report the results of analysis of the obtained *attack–defense tree* (ADTree) model using some of the recent developments in the field.

© Springer Nature Switzerland AG 2019
M. Albanese et al. (Eds.): GraMSec 2019, LNCS 11720, pp. 95–125, 2019.
https://doi.org/10.1007/978-3-030-36537-0_6

Our Objectives. The objective of the work presented in this report is threefold. First, we thoroughly analyze a tampering with a power meter scenario which is of high importance for the industrial community. The goal is to identify optimal attacks and optimal defender's strategies, while taking various optimality criteria into account. Second, we test in practice some recently proposed techniques for quantitative analysis of ADTrees: the one from [25], focusing on the correctness of the quantitative analysis of trees with repeated actions, the one from [13], addressing the problem of multi-parameter evaluation on such trees, and the one from [24], which relies on linear integer programming (ILP) and aims at selecting optimal sets of countermeasures. Third, we develop and validate a tool, called OSEAD – *Optimal Strategies Extractor for Attack–Defense trees* – supporting an automated handling of the techniques from [24,25], and [13].

Related Studies. ADTrees have already been used in the past to perform practical studies of security scenarios. The security of ATM machines was analyzed in [14]. The main difference between [14] and the current study is that the former focuses on the modeling aspects only, i.e., it does not involve any quantitative analysis. In [6], an RFID-based management system has been analyzed. This work resulted in a list of guidelines describing how to carry out a case study involving the ADTree modeling and its quantitative analysis. These guidelines were respected in our power meter study. However, [6] concentrates on analysis wrt single parameter, and it uses the classical bottom-up approach, sketched in [38] and formalized in [30] and [21], which is not well-suited for trees with repeated basic actions. In the current work, we employ novel, recently developed techniques for quantitative analysis of ADTrees with repeated basic actions, and use and validate a tool that we developed to support an automated handling of these techniques. Unlike in [6] and [14], the power meter ADTree is based on an attack tree developed by industry, so the people performing the analysis are not the sole authors of the analyzed model.

Paper Structure. The ADTree for tampering with the power meter is described in Sect. 2. The parameters of interest for our study, their values, and the techniques we employ to identify optimal attacks and optimal defender's strategies are presented in Sect. 3. The results of the actual analysis are discussed in Sect. 4, and Sect. 5 is devoted to the OSEAD tool. We conclude in Sect. 6.

2   Tampering with a Power Meter Scenario

We start with presenting the scope of this study, which includes the profiles of the actors involved, a specification of the analyzed system, and the modeling technique that we use, in Sect. 2.1. Then, in Sect. 2.2, we explain the scenario analyzed in this study.

2.1   The Set-Up

We consider a fifth year student of an engineering school, whom we will name *Antoine*, who is renting an apartment where he needs to pay for the electricity consumption. Antoine would like to lower his electricity bill and he decided to achieve this by reconfiguring the power meter in his apartment. In this study, Antoine plays a role of an *attacker* and his opponent, i.e., a *defender*, is the electricity provider. The meter under study is equipped with an optical port that allows a user to connect to the meter using an optical probe. Illustrations of a meter (Fig. 12) and an optical probe (Fig. 13) can be found in Appendix A.

To describe possible ways in which Antoine may reprogram the meter, we use attack–defense trees [21]. An attack–defense tree (ADTree) is a graphical security model representing how an attacker may attack the analyzed system and how the defender may counter these attacks. The main building blocks of an ADTree are the *labels* of its nodes, two types of *refinements* – OR and AND – and the *countermeasures*. The labels describe the goals that the attacker or the defender want to achieve. The refinements allow to decompose these goals into simpler subgoals (refined nodes) and basic actions (non-refined nodes). To achieve the goal of an OR node, the corresponding agent needs to achieve the goal of at least one of its child nodes. The goal of an AND node is achieved if the goals of all of its children are achieved. A countermeasure attached to a node of one agent represents how the node's goal may be overcome by the other agent. This means that a node of the attacker can be countered by a node of the defender, which in turn can be counterattacked by a node of the attacker, and so on. In addition, a countermeasure can also be decomposed using OR and AND refinements.

ADTrees extend the well-known model of attack trees [38] widely used in industry [10, 11, 33]. The starting point of our analysis was the scenario and the attack tree described in Section 2.3 of [33]. We complemented this tree with additional attacks, and added possible countermeasures that we identified based on [9,31], and [40]. The resulting ADTree contains 68 nodes, 5 repeated basic actions of the attacker and 3 repeated basic actions of the defender. Following the convention established in [8] and [25], we assume that nodes having the same label represent exactly the same instance of an action. They thus model that executing some action may contribute to several different attacks or defenses.

2.2 The Scenario

In order to reconfigure his power meter via optical port, Antoine has to have physical access to the power meter and reconfigure it using appropriate software tools. Since the power meter is located in the apartment where Antoine lives, we assume that accessing the power meter is a basic action, i.e., the corresponding node is not refined. In order to reconfigure the power meter with the help of software, we have identified the following three sub-scenarios that Antoine can follow, taking into account his knowledge, capabilities, and financial profile:

The *do it yourself* approach – Antoine reconfigures the meter himself by using unauthorized software tools (Figs. 2, 3, 4, 5 and 6),
The *social engineering* approach – Antoine social engineers a technician employed by the electricity provider to reconfigure the power meter for him using authorized software tools (Fig. 7),

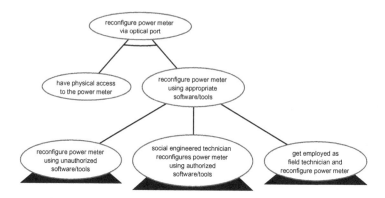

Fig. 1. How to reconfigure the power meter – a high level view (Color figure online)

The **get employed** approach – Antoine gets employed by the electricity provider as a field technician to gain access to the authorized tools and to be able to reconfigure the meter himself (Fig. 8).

This high-level view of the analyzed scenario is presented by the tree from Fig. 1, where the black triangles illustrate subtrees presented in further figures. In this work, we use the standard graphical notation for ADTrees: red and green nodes represent attacker's and defender's goals, respectively, and an arc is used to denote the AND refinement. In the rest of this section, we detail the three approaches considered by Antoine.

The *do it yourself* approach

To reconfigure the power meter by himself, Antoine needs to obtain unauthorized software and tools, use optical probe to establish connection with the meter via its optical port, and finally reconfigure the meter using unauthorized software. He can find and download unauthorized software from the Internet. As for the optical probe, he can buy it or make it himself. The corresponding tree is given in Fig. 2.

Establishing connection to the meter via its optical port might be secured by password authentication. Also, independently of whether a password-based protection is implemented or not, an authentication could be required before the power consumption configuration can be modified. These two possible counter-measures are depicted with the green nodes in Fig. 2.

If the connection to the power meter was protected by a password, Antoine could still reach his goal if he was able to authenticate using the correct credentials. To do so, he would need to obtain the credentials and enter them to the power meter while authenticating, as visualized in Fig. 3. The power meter credentials could be obtained by

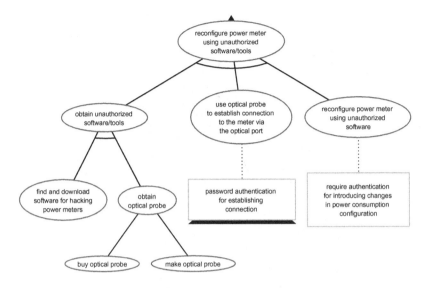

Fig. 2. The *do it yourself* approach (Color figure online)

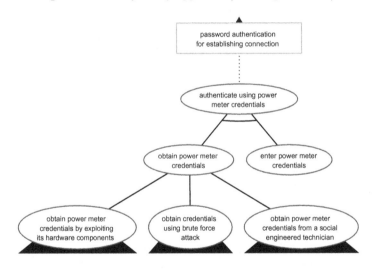

Fig. 3. Overcoming the password-based authentication

- exploiting the hardware components of the power meter (Fig. 4),
- performing a brute force attack (Fig. 5), or
- social engineering a technician working for the power provider (Fig. 6).

Extracting credentials from the power meter hardware components, illustrated in Fig. 4, can be achieved in two ways: either by extracting them from a data dump or by spying on communication between the hardware components. To extract the credentials from the data dump, the dump needs to be made, the location where encrypted credentials are placed in the dump needs to be

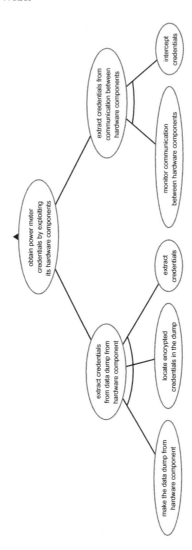

Fig. 4. Obtain credentials from hardware components

identified, and finally the credentials need to be extracted from the encrypted dump. To extract the credentials from the communication between the hardware components, the communication needs to be monitored and the credentials need to be intercepted. In this study, we assume that during the communication between the hardware components, the data are sent unencrypted.

A brute force attack is illustrated in Fig. 5. It makes use of software for hacking power meters (in our scenario, this is exactly the same software as the one used by the attacker to reconfigure power meter). An off-line brute force attack using tools like Ophcrack [34], John the Ripper [35], or hashcat [18], can be

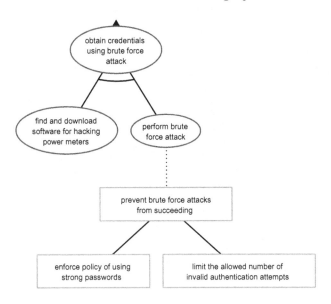

Fig. 5. Obtain credentials by brute force attack

prevented if a strong password is used. To make an on-line cracking impossible, the number of possible invalid authentication attempts could be limited.

Finally, credentials could also be obtained by social engineering a technician, as depicted in Fig. 6. To do so, a suitable technician would need to be selected and social engineered. A social engineering attack would require to assemble background information on employees of the energy provider and select one who would fall into the social engineering attack to reveal the credentials. Antoine could obtain the background knowledge on employees by searching on the Internet, diving into dumpster and looking for relevant documents and physical artefacts, or by infiltrating the energy provider. To infiltrate the energy provider, Antoine could get hired as an intern student and then collect information by exchanging gossips with the company employees. The following policies could be enforced by the company to prevent access to the background information about its employees:

- a policy to minimize the Internet disclosure,
- a policy to minimize the leakage of physical documents and artefacts,
- a policy of performing thorough background check before hiring new employees.

Once the right social engineering target is selected, the attack itself consists in bribing, coercing or tricking the technician so that they reveal the power meter credentials. The tricking attack could be prevented by a security training during which the personnel is made aware of popular social engineering tricks.

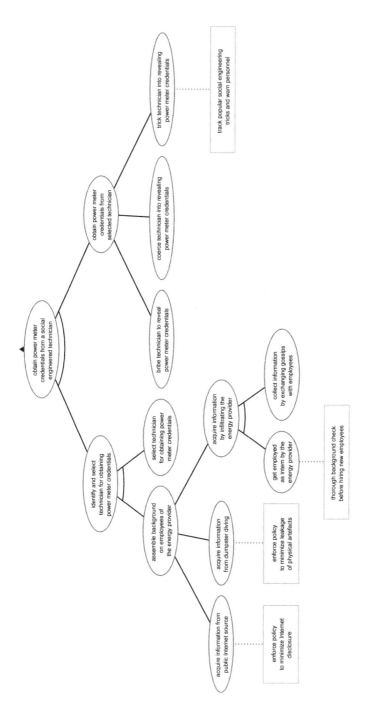

Fig. 6. Obtain credentials by social engineering a technician

The *social engineering* approach

Instead of attacking by himself, Antoine can social engineer a technician, so that they reconfigure the power meter for him, as modeled in Fig. 7. To perform the social engineering, a suitable technician who would reconfigure the power meter needs to be identified and Antoine needs to convince them to reconfigure the meter. Identification of the suitable social engineering target is performed in exactly the same way as in the *do it yourself* approach, by assembling relevant background knowledge on employees. Once identified, the technician who will reconfigure the power meter is selected. To persuade the technician to reconfigure the power meter, Antoine can bribe or coerce them.

Fig. 7. The *social engineering* approach

The *get employed* approach

Antoine can also get hired by the power provider company to be officially able to reconfigure power meters. To do so, he needs to get employed as a field technician and then reconfigure his power meter using authorized software provided by the company to its technicians. Performing thorough background check on future employees would mitigate this attack, as it was the case in the two previous approaches. The get employed attack is illustrated in Fig. 8.

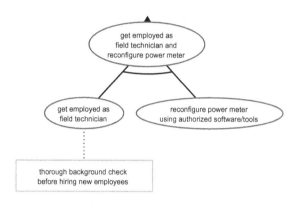

Fig. 8. The *get employed* approach

3 Quantitative Analysis of the Tampering Scenario

The first objective of this case study is to analyze the scenario described in Sect. 2. This includes enumeration of all possible attacks, identification of those that are optimal from the point of view of the attacker, as well as pinpointing the countermeasures that offer the best protection to the analyzed system. A plethora of algorithms exist to perform quantitative analysis of ADTrees, e.g., [5,16,21,22,24,25]. However, some of them may produce incorrect results when applied to ADTrees where an action may contribute to several attacks. In this study, we thus focus on the methods developed in [13,25] which have been especially developed to deal with ADTrees containing such repeated basic actions, and an adaptation of the methods of [24] applicable to such trees.

In what follows, we will use the notion of an *attack* which we define as *a minimal set of basic actions of the attacker the execution of which achieves the goal of the root node of the ADTree, under some fixed behavior of the defender.* By *defender's strategy*, we understand a *set of countermeasures* that the defender can implement to secure the system.

In Sect. 3.1, we explain the problems that are of interest for our study. Then, in Sect. 3.2, we provide clear definitions of the parameters that we have used. In Sect. 3.3, we present the input values that we have used and we explain how they have been estimated. Finally, in Sect. 3.4, we discuss some issues related to the reliability of the input values and the computation methods used.

3.1 The Problems of Interest

The three types of optimization problems that we tackle in this study are:

- selection of attacks optimal wrt one parameter,
- selection of attacks optimal wrt several parameters,
- selection of the defender's strategy optimal from the point of view of their resources and objective.

The parameters (often also called *attributes* in the context of ADTrees [22]) used in this work are *cost, time, probability,* and *special skills*. Before we present their meaning in Sect. 3.2, we provide generic, i.e., parameter independent, explanation of the problems that we address.

First, we are interested in selecting the *best* attacks, taking only one parameter into account. Here 'best' means optimal wrt the parameter considered. An optimal attack may thus be a one that minimizes *time* or *cost*, or maximizes *probability* of success. The selection of optimal attacks is performed according to the method proposed in [25], which ensures that the computations on ADTrees with repeated labels yield correct results, contrary to the classical bottom-up approach, used for instance in [22]. This method also allows to rank possible attacks, and return k most optimal ones, where $k \in \mathbb{N}$ can be specified by the user.

Second, it is well known that considering one parameter at a time may not be sufficient to decide which attack is the best (or the worst, if the scenario is analyzed from the defender's point of view). A possible approach to address this problem is to identify *Pareto optimal attacks*, i.e., attacks that are undominated by others, when several parameters are regarded simultaneously. In our study, we employ the framework developed in [13] to select the Pareto optimal attacks extracted from an ADTree with repeated labels. The solution proposed in [13] relies on so called *Pareto attribute domains* that allow for combining a number of parameters into a single one, and therefore make it possible to perform a multi-objective analysis using the same tools as in the case of a single parameter.

Finally, the solutions to the two problems presented above depend on which countermeasures have been implemented to protect the system. Instead of considering all possible cases, it is thus wise to determine a strategy which from the point of view of the defender is optimal. In an ideal world, the best would be to implement all possible countermeasures. However, in real-life, this is not possible, because the budget of the defender is limited. Also, it does not make sense to implement countermeasures which do not add more security wrt the ones already in place. To select an optimal strategy of the defender, we use the method relying on integer linear programming proposed in [24], adapted to ADTrees with repeated labels. It allows for selecting one of the two criteria – *maximizing the number of countered attacks* (coverage problem) or *maximizing the necessary investment of the attacker* (investment problem) – and selecting the defender's strategy which optimizes the chosen criterion and is compatible with the defender's budget.

3.2 The Parameters Used

We now describe the parameters that we have used in our study to address the problems presented in Sect. 3.1. We give an intuitive meaning to each parameter, provide the set of values that they can attain, and explain how the parameter's value of an attack is computed from the values corresponding to the attack's components.

Cost (min, +). The first parameter of interest is the *monetary investment* necessary to implement an attack (or a defender's strategy). To express it, we use non-negative real numbers representing the necessary investment in euro. The actions that are too expensive to be executed are assigned the value of $+\infty$. When optimizing the cost parameter, we are interested in attacks (or sets of defenses) yielding minimal monetary investment for the attacker (resp. defender). Thus, to compute the cost of an attack (resp. set of defenses), we *sum* the values of the basic actions composing it. To select the best attack (resp. defender's strategy), we then choose the one with *minimal* value.

Time (min, max). Since Antoine would like to lower his electricity bill as soon as possible, the *time that an attack would take* is an important parameter to consider. The following scale is used to express time values:

- *Instantaneous* (0): can be performed by the actor in less than a minute.
- *Quick* (10): can be performed by the actor in less than an hour, but not less than a minute.
- *Slow* (10^2): can be performed by the actor in less than a week, but not less than an hour.
- *Very slow* (10^3): can be performed by the actor in less than six months, but not less than a week.
- *Extremely slow* (10^4): can be performed by the actor within a human lifetime, but not less than six months.
- *Impossible* ($+\infty$): not doable within a human lifetime.

Since this scale is discrete, it is reasonable to assume that the time necessary to perform an attack is the *maximum* value over the time values of its composing actions. As in the case of cost, we are interested in *minimizing* the time necessary to attack the system, thus we select the attack which requires *minimal* time.

Success probability (max, ×). Attacks that are very cheap or very fast are useless if their probability of succeeding is negligible. Here, we are thus interested in what is the probability that, if executed, an attack will be successful. The probability of an action is a value from the interval $[0, 1]$, and the probability of an attack is computed by multiplying[1] When optimizing, we select the attack with

---

[1] Using multiplication implies that attack components are considered to be independent. This is a known drawback of the classical bottom-up propagation of probability values in attack tree-based models. See Sect. 3.4 for a discussion on this and related issues.

the *maximal* probability of success, because this is the attack that a reasonable attacker would prefer.

The remaining three parameters assess the level of special skills – cybersecurity, technical, and social – that is necessary to be able to perform an action successfully. In all three cases, the skill level necessary to perform an attack is defined as the *maximum* amongst the skill levels necessary to perform its components. By optimal, we mean an attack requiring *minimal* skill level.

Cybersecurity skills level (min, max). Some of the actions considered in our scenario may require specific *expertise regarding cybersecurity*. We distinguish five levels of such expertise:

- *None* (0): no cybersecurity-related skills required.
- *Basic* (1): requires basic cybersecurity knowledge and skills.
- *Advanced* (2): requires employing advanced cybersecurity-related skills, e.g., executing a man in the middle attack on a protocol.
- *Expert* (3): requires employing cybersecurity-related skills available to few experts, e.g., return-oriented programming or fault attack on AES.
- *Impossible* ($+\infty$): beyond the known capability of today's human beings.

Technical skills level (min, max). Similarly to cybersecurity skills, some actions may require some *technical expertise*. Here again, we distinguish five levels:

- *None* (0): no technical skills required.
- *Basic* (1): requires basic technical skills, e.g., finding information online.
- *Advanced* (2): requires advanced technical skills, available for graduates of technical vocational schools.
- *Expert* (3): requires technical skills available to experienced engineers.
- *Impossible* ($+\infty$): beyond the known capability of today's human beings.

Social skills level (min, max). Finally, since some attacks in our scenario rely strongly on social engineering, we are also interested in *social skills* necessary to perform the considered actions. The five levels of social skills are defined as follows:

- *None* (0): does not involve social interactions.
- *Basic* (1): requires basic social interactions, e.g., obtaining information via a conversation.
- *Advanced* (2): requires convincing or tricking someone into doing something they would not do otherwise.
- *Expert* (3): requires convincing or tricking someone into doing something punishable by law.
- *Impossible* ($+\infty$): beyond the known capability of today's human beings.

3.3 Estimation of Input Values

We solve the optimization problems presented in Sect. 3.1 using frameworks developed in [24,25] and [13]. The underlying algorithms take as input an ADTree as well as parameter values assigned to its basic actions.

Table 1 gathers the basic actions of the defender and gives their cost. The values of the defender's cost represent the investment that the electricity provider needs to make to hire security experts who will advise the company on potential threats and suitable countermeasures against them, organize meetings where the decisions on policies to be implemented will be taken, put in place improved software or hardware solutions, for instance those allowing more secure authentication, and remunerate its personnel for performing specific activities, such as background checks before hiring new employees.

Table 1. Cost of basic actions of the defender

Basic action	Cost
d_1 = enforce policy of using strong passwords	11600
d_2 = enforce policy to minimize Internet disclosure	9600
d_3 = enforce policy to minimize leakage of physical artefacts	9600
d_4 = limit the number of possible invalid authentication attempts	11600
d_5 = password authentication for establishing connection	13600
d_6 = require authentication for introducing changes in power consumption configuration	13600
d_7 = thorough background check before hiring new employees	320
d_8 = track popular social engineering attacks and warn personnel	1500

The values of basic actions of the attacker that we have used in this study are given in Table 2. They represent a consensus reached as a result of the following procedure. Seven independent participants, whose profiles correspond to the expertise of Antoine, were involved in the values' estimation. The participants were given a document describing the scenario and the ADTree from Sect. 2. They had access to the Internet and relevant materials, including [9,33,40]. Each participant estimated the values for all six parameters at every basic action present in the tree. Unsurprisingly, some of the values were not consistent amongst different participants. A semi-automatic procedure has thus been used to extract a single value for each parameter at every basic action:

- for the parameters different than *probability*: if all (but one) amongst the seven values were the same, this value was retained,
- for the *probability* parameter, a simple average over seven values was computed,
- for the cases that do not fall into any of the above items, the retained value is the result of a discussion between the two authors of this paper,

Table 2. Parameter values for basic actions of the attacker

Basic action	Cost	Time	Prob	Cyber	Tech	Social
acquire information from dumpster diving	0	1000	0.2	0	0	0
acquire information from public Internet source	0	100	0.79	0	1	0
bribe technician to reconfigure the power meter	500	10	0.52	0	0	3
bribe technician to reveal power meter credentials	300	10	0.5	0	0	2
buy optical probe	71.2	100	1	0	1	0
coerce technician into reconfiguring the power meter	0	100	0.3	0	0	3
coerce technician into revealing power meter credentials	0	100	0.33	0	0	3
collect information by exchanging gossips with employees	0	1000	0.46	0	0	1
enter power meter credentials	0	0	0.99	0	0	0
extract credentials	0	10	0.56	0	1	0
make the data dump from hardware component	0	100	0.73	1	3	0
find and download software for hacking power meters	0	10	0.9	1	1	0
get employed as field technician	0	1000	0.48	0	2	1
get employed as intern by the energy provider	0	1000	0.52	0	1	1
have physical access to the power meter	0	0	1	0	0	0
intercept credentials	0	0	0.62	2	1	0
locate encrypted credentials in the dump	0	100	0.6	2	2	0
make optical probe	14	100	0.41	0	2	0
monitor communication between hardware components	0	100	0.5	1	2	0
perform brute force attack	0	100	0.65	1	2	0
provide power meter credentials	0	0	1	0	0	0
reconfigure power meter using authorized software/tools	0	10	0.94	0	1	0
reconfigure power meter using unauthorized software	0	10	0.75	0	2	0
select technician for obtaining power meter credentials	0	100	1	0	0	0
select technician for reconfiguring power meter	0	100	1	0	0	0
technician reconfigures power meter using authorized software/tools	0	10	1	0	0	0
trick technician into revealing power meter credentials	0	100	0.24	0	0	2
use optical probe to establish connection to the meter via the optical port	0	10	0.95	0	1	0

– finally, in the case of strong disagreement, the author of the analyzed ADTree who, amongst the seven participants, knows the best the optical meter technology, had the decisive power.

The estimation of values took one hour to each participant, on average. The consensus discussions lasted for three hours, in total.

3.4   Debating on the Reliability of the Computation framework

Quantifying security is a highly disputable exercise. The reliability of the obtained results depends on the quality of the employed input values and on the suitability of the functions used to perform computations. Despite a great effort of the academic and the industrial communities, numerous underlying issues still remain unsolved. In this section, we debate on drawbacks that we met while performing this study, some of which we have not necessarily managed to overcome.

The quantitative analysis of graphical security models relies on numerical inputs whose exact values can almost never be provided. Their estimation is a difficult task that requires a thorough understanding of

– the parameters employed,
– the meaning of the basic actions present in the tree,
– the attacker's and defender's profiles and knowledge.

In practice, this estimation is very subjective, as it relies to a great extent on the modeler's expertise. In real-life, input values are usually based on historical data, statistics, information gathered from surveys or open sources, e.g., Internet. Such inputs inevitably carry some uncertainty about the values, and this uncertainty propagates during the computations and is accumulated in the final result of the analysis. While there is no established methodology for determining the best approximations of the actual values of the parameters under consideration, we believe that a reasonable estimates can still be obtained, if provided in collaboration with experts in the respective domains. Several industry practitioners performing security and risk analysis on a daily basis, that we had an opportunity to work with, suggest to follow a couple of simple rules.

– Finding a consensus through a discussion usually results in numbers that are more accurate than standard composite values, e.g., the average. People providing inputs might have misunderstood the significance of a parameter or the meaning of an action, thus their values might be inconsistent. Computing a simple average over such values is meaningless. A discussion allows to identify such misunderstandings and results in an estimate reflecting the reality properly.
– If a discrete scale is used, an odd number of possible values, such as *low-medium-high*, should be avoided. People having problems with deciding on the most suitable value, for instance due to the lack of knowledge, often tend to go for the middle value, because it seems to be the most neutral alternative. However, if numerous attacks get the same value, their ranking and thus a selection of the optimal ones become impossible.

– A way of taking the knowledge of the value providers into account is to complement the parameter value with the information on how certain the provider is about this value. Such an approach has, for instance, been used in the case study described in [6], where a *confidence* level was used in addition to the actual values of the parameters of interest. The confidence level plays a role of a weight, allowing to give more importance to values with high confidence (usually provided by experts) compared to those with low confidence (probably coming from less knowledgeable participants).

Note that, in our study, we decided not to use the confidence level, because our value providers had exactly the same profile as our potential attacker Antoine. We thus assumed that their estimates would be consistent with the estimates (and thus indirectly with decisions) that Antoine would make.

Another factor possibly undermining the pertinence of the quantitative analysis of security are the computations performed on the input values during the analysis. We illustrate this issue on the examples of *probability* and *risk* metrics. An arguable but commonly used operator in the context of attack tree analysis is the multiplication employed to propagate the probability values at AND nodes in a bottom-up fashion. Using multiplication implies that attack components are considered to be independent, which is rarely the case in reality. This means that, even if the input values are correct, the probability computation might introduce some error or inaccuracy to the final result. To overcome this known drawback of the classical bottom-up propagation, some more advanced methods for computing attacks' probability have been proposed in the literature. Their weakness however lies in the fact that they often require sophisticated inputs, such as conditional probability tables [23] or probability distributions [3], instead of simply probability points. The interested reader is referred to Section 7 of [41] for a description of existing probabilistic frameworks for attack tree-based analysis. Another example highlighting the importance, but at the same time the difficulty of quantifying security is the *risk* metrics. Various formulas for risk exist. In [37], the authors state that the standard way of defining risk is 'the likelihood of an incident and its consequences for an asset', with all the words used having some specified meaning. This definition is used for instance in the French risk analysis method *EBIOS* [1]. It relies on two factors only, but other definitions are possible. In [12], risk has been defined in terms of cost, probability, and impact. For a discussion on possible three-factor and many-factor risk measure definitions see Chapter 11 of [37] and references therein. On the one hand, the fact that there are many risk metrics definitions can be seen as a positive thing, because it allows the expert to select the one that is most suitable in a specific analysis context or wrt the available input values. On the other hand, however, different risk formulas will provide different results, so it might be unclear which risk formalization should be used in which case.

To conclude this discussion section, we would like to stress that graphical security models are not the silver bullet for the risk assessment process, and that their role is to accompany other threat and risk analysis approaches, such as penetration testing, red teaming, standardized ISO 27XXX-compatible methods, e.g., [1, 29], etc. Each of these methods focuses on different types of attacks and different security problems, so it is worthwhile to combine them in order to get the most complete and full-fledged results.

4 Optimal Strategies for the Attacker and the Defender

We now present the results of the power meter tampering scenario analysis. We begin, in Sect. 4.1, by determining sets of countermeasures that the defender can implement under specified budget and that are optimal wrt a given criterion (coverage or attacker's investment). For some of these sets, we then perform a *what-if* analysis: if a given strategy of the defender is implemented, what are the attacks optimal wrt one (Sect. 4.2) or many (Sect. 4.3) parameters? Our objective is to verify whether an attacker having a profile of Antoine would be able to launch a successful attack on its power meter. Due to space restrictions, this section summarizes some main observations drawn from our analysis. The files presenting the raw data and all the obtained results are available at https:// people.irisa.fr/Wojciech.Widel/studies/meter_study.zip.

Performing this analysis is a laborious task, because the underlying algorithms are complex [13, 24, 25] and our tree is too big to be analyzed manually. To obtain the results presented in this section, we have used the OSEAD tool that supports the techniques proposed in [13, 24], and [25]. We postpone the technical description of OSEAD to Sect. 5.

4.1 Selection of Optimal Sets of Countermeasures

The choice of an optimal strategy for the defender depends on the budget that they have at their disposal, and on the optimization problem of interest. In our study, we consider a small, local electricity provider, and we thus analyze three possible values for the defender's budget: 20000, 30000, and 40000 euros. Table 3 presents optimal strategies for a defender interested in maximizing the number of prevented attacks (coverage problem) and another one focused on maximizing the necessary investment of the attacker necessary to achieve his objective (investment problem). They have been obtained using the OSEAD tool.

Requiring authentication for introducing changes in power consumption con-figuration (d_6) and *performing thorough background check before hiring new employees* (d_7) is an optimal strategy for a defender interested in covering a maximal number of possible attacks and having the budget of 20000 euros. We denote this strategy by D_1. Under the same budget, but with the goal of max-imizing the necessary investment of the attacker in mind, the optimal behavior of the defender would be to *enforce policy to minimize Internet disclosure* (d_2), *enforce policy to minimize leakage of physical artefacts* (d_3) and *perform thor-ough background check before hiring new employees* (d_7). This ensures that the minimal necessary investment of the attacker into achieving the root goal is 14. This means, in particular, that the execution of the three actions prevents all the attacks having cost of 0 euros.

The other two strategies that we consider are D_2 which corresponds to D_1 extended with the action of *enforcing policy to minimize Internet disclosure* (d_2), and D_3 consisting of *enforcing policy to minimize Internet disclosure* (d_2), *enforcing policy to minimize leakage of physical artefacts* (d_3), *performing thorough background check before hiring new employees* (d_7), and *tracking popular social engineering attacks and warning personnel* (d_8). These strategies are optimal for a defender having 30000 euros, and interested in the coverage problem (D_2) and the attacker's investment problem (D_3), respectively.

Finally, a defender having 40000 euros is able to fully secure the analyzed system, by implementing countermeasures d_2, d_3, d_6, and d_7. Due to space restrictions, we refer the reader to Table 1 for their meaning.

Table 3. Optimal strategies of the defender

Defender's budget	Coverage problem		Investment problem	
	Optimal strategy	Prevented/ preventable	Optimal strategy	Necessary attacker's investment
20000	$D_1 = \{d_6, d_7\}$	29/33	$\{d_2, d_3, d_7\}$	14
30000	$D_2 = \{d_2, d_6, d_7\}$	31/33	$D_3 = \{d_2, d_3, d_7, d_8\}$	14
40000	$\{d_2, d_3, d_6, d_7\}$	33/33	$\{d_2, d_3, d_6, d_7\}$	$+\infty$

For the rest of our study, we retain the strategies D_1, D_2, and D_3 and look for optimal attacks in the case when one of these strategies is implemented by the defender.

4.2 Attacks Optimizing Single Parameter

For determining attacks optimal wrt to one parameter, OSEAD first extracts the attacks from the model (attacks in the sense of *set semantics* defined in [25]), and then computes the values of the parameter corresponding to each of them. In total, there are 33 attacks in the studied scenario, and their list is available at https://people.irisa.fr/Wojciech.Widel/studies/meter_attacks.txt. Whether an attack is successful or not depends on the countermeasures that are implemented by the defender. The attacks of interest for us are those that are not countered by at least one of the three defender's strategies D_1, D_2 or D_3. There are twelve such attacks, and they are presented in Table 4.

By analyzing the data gathered in Table 4, we notice that if the defender decides to implement one of the strategies D_1 or D_2, Antoine will be able to succeed only by executing some of the attacks from the *social engineering* approach. If the strategy D_3 is implemented, then the only possible attacks are those from the *do it yourself* approach.

Once the values corresponding to the attacks are obtained, OSEAD returns the optimal ones. We list them in Table 5. This table can be used to check whether

Table 4. Some of the attacks available to Antoine

Attacking approach: do it yourself (Y); social engineering (S); get employed (E)	S	S	S	S	Y	Y	Y	Y	Y	Y	Y	Y
Basic action	A_1	A_2	A_3	A_4	A_5	A_6	A_7	A_8	A_9	A_{10}	A_{11}	A_{12}
acquire information from dumpster diving		✓		✓								
acquire information from public Internet source	✓		✓									
bribe technician to reconfigure the power meter			✓	✓								
bribe technician to reveal power meter credentials												
buy optical probe									✓	✓	✓	✓
coerce technician into reconfiguring the power meter	✓	✓										
coerce technician into revealing power meter credentials												
collect information by exchanging gossips with employees												
enter power meter credentials					✓		✓	✓		✓	✓	✓
extract credentials							✓	✓				
find and download software for hacking power meters					✓	✓	✓	✓	✓	✓	✓	✓
get employed as field technician												
get employed as intern by the energy provider												
have physical access to the power meter	✓	✓	✓	✓	✓	✓	✓	✓	✓	✓	✓	✓
intercept credentials							✓	✓				
locate encrypted credentials in the dump							✓		✓			
make optical probe					✓	✓	✓	✓				
make the data dump from hardware component							✓		✓			
monitor communication between hardware components							✓		✓			
perform brute force attack					✓							✓
provide power meter credentials												
reconfigure power meter using authorized software/tools												
reconfigure power meter using unauthorized software					✓	✓	✓	✓	✓	✓	✓	✓
select technician for obtaining power meter credentials												
select technician for reconfiguring power meter	✓	✓	✓	✓								
technician reconfigures power meter using authorized software/tools	✓	✓	✓	✓								
trick technician into revealing power meter credentials												
use optical probe to establish connection to the meter via the optical port					✓	✓	✓	✓	✓	✓	✓	✓
Defender's strategy under which the attack is successful	D_1	D_1, D_2	D_1	D_1, D_2	D_3	D_3	D_3	D_3	D_3	D_3	D_3	D_3

an attacker of interest would be able to launch a successful attack. We recall that Antoine is a fifth year student of an engineering school. We assume that he has advanced technical skills, but he has only basic knowledge of cybersecurity. As most of students, he is not rich, but he can manage his time availability freely.

On the one hand, since the cost aspect is of the highest priority for Antoine, he would analyze the attacks optimal wrt to this parameter first, see the *Cost* column in Table 5. The preference is given to attack A_2 which consists of *having physical access to the power meter, acquiring information from dumpster diving, selecting technician for reconfiguring power meter, coercing technician into reconfiguring power meter* and the *technician reconfiguring power meter using authorized software/tools*. While this attack is optimal from the point of view of *cost* and all the three *skills levels* under strategies D_1 and D_2, it would require from Antoine to force someone to perform an action punishable by law. Also, A_2 is prevented by the strategy D_3. Indeed, implementation of D_3 counters all the attacks from the *social engineering* approach.

On the other hand, D_3 does not secure the meter from any attack in the *do it yourself* approach. An interesting attack within this approach is A_6, consisting of *having physical access to the power meter, making optical probe, finding and downloading software for hacking power meters, using optical probe to establish connection to the meter via the optical port*, and *reconfiguring power meter using unauthorized software*. Note that A_6 corresponds to the profile of Antoine, from the point of view of his resources and skills. Its only drawback is that its probability of success is quite low – only 0.26, as can be seen in Table 6.

Thanks to Table 5, we can also study the impact of the implemented countermeasures on the attacks available to the attacker. Upgrading the system's protection from D_1 do D_2 (by *enforcing policy to minimize Internet disclosure*) at the cost of 9600 euros (see Table 1) is not worthwhile if the defender considers cheap attacks to be the most tempting for the attacker – the attack A_2 achieves the root goal under both strategies D_1 and D_2. However, if the defender aims at making the attacker less likely to succeed, then this investment is beneficial, as it lowers the attacker's success probability from 0.41 (for attack A_3 which would not work under D_2) to 0.10 (for A_4 that still works when D_2 is implemented).

4.3 Attacks Optimizing Several Parameters

Unfortunately, for every attack listed in Table 5, i.e., optimal wrt to some given parameter, there always exists another one that is better from the point of view of another parameter. To overcome this problem, we are now looking for Pareto optimal attacks, i.e., attacks that are not dominated by another one, while taking all six parameters into account simultaneously.

The Pareto optimal attacks are presented in Table 6, along with the values corresponding to their execution. Observe that under strategies D_1 or D_2, all of the attacks available to Antoine are Pareto optimal, including the attack A_2 discussed in the previous section. If the strategy D_3 is implemented by the defender, there exist eight possible attacks that achieve the root goal, but only two of them are Pareto optimal, namely A_6 and A_9. It is crucial to notice that

Table 5. Attacks optimal wrt a single parameter and their values

Defender's strategy	Attacks optimal wrt					
	Cost	Time	Prob	Cyber	Tech	Social
D_1	A_1, A_2	A_1, A_3	A_3	$A_1, A_2,$ A_3, A_4	A_2, A_4	$A_1, A_2,$ A_3, A_4
Optimal value	0	100	0.41	0	0	3
D_2	A_2	A_2, A_4	A_4	A_2, A_4	A_2, A_4	A_2, A_4
Optimal value	0	1000	0.10	0	0	3
D_3	$A_5, A_6,$ A_7, A_8	$A_5 - A_{12}$	A_9	$A_5, A_6,$ A_9, A_{12}	$A_5, A_6,$ $A_8, A_9,$ A_{11}, A_{12}	$A_5 - A_{12}$
Optimal value	14	100	0.64	1	2	0

A_9 is a very interesting attack. It is almost the same as A_6, except that it involves *buying optical probe* instead of *making it*. Attack A_9 is optimal wrt to all parameters, except *cost*. However, when checking its *cost* value, one realizes that the investment necessary to perform it (71.2 euros) would probably be acceptable for Antoine. The greatest advantage of A_9 is that its success probability (0.64) is significantly higher than that of A_6 (0.26).

Table 6. Pareto optimal attacks and their values for: *cost* (*c*), *time* (*t*), *prob* (*pb*), *cyber skills* (*cs*), *tech. skills* (*ts*), and *social skills* (*ss*)

Defender's strategy	Pareto optimal attacks	Values (c, t, pb, cs, ts, ss)
D_1	A_1	(0, 100, 0.24, 0, 1, 3)
	A_2	(0, 1000, 0.06, 0, 0, 3)
	A_3	(500, 100, 0.41, 0, 1, 3)
	A_4	(500, 1000, 0.10, 0, 0, 3)
D_2	A_2	(0, 1000, 0.06, 0, 0, 3)
	A_4	(500, 1000, 0.10, 0, 0, 3)
D_3	A_6	(14.0, 100, 0.26, 1, 2, 0)
	A_9	(71.2, 100, 0.64, 1, 2, 0)

The importance of the multi-parameter analysis is further illustrated by two facts. First, securing the system in a way that maximizes the necessary investment of the attacker, by implementing D_3, not only leaves the system vulnerable to more attacks than it is the case for the coverage problem (eight attacks versus two or four, see last row of Table 4), but also allows the attacker to execute attack A_9, which has a high probability of succeeding. Second, when the defender implements strategy D_3, the attack A_6 is among the cheapest ones, and the attack

A_9 is the optimal one wrt the probability. When we analyze the scenario taking only one of these parameters into consideration, we overlook one of these two attacks. But both of them are Pareto optimal, and as such, both can be considered equally appealing for the attacker.

5 The OSEAD Tool

We now present the OSEAD tool – *Optimal Strategies Extractor for Attack–Defense trees* – that we used for performing analysis of Sect. 4. We describe its features in Sect. 5.1, its performance in Sect. 5.2, and give some implementation details in Sect. 5.3.

5.1 OSEAD from the User's Perspective

OSEAD allows its users to benefit from the theoretical developments of [13,24,25] in a simple and intuitive way. Users operate the tool in a step-by-step manner, via a graphical interface illustrated in Fig. 9. The first step is to provide a file storing the structure of the ADTree of interest, which is an XML file produced by ADTool [17], well-known software for creating ADTrees. Furthermore, should the user want to analyze an attack tree created with the help of ATCalc [2] or ATE [4], the output files of these tools can be easily transformed into an ADTool-like XML file with the help of ATTop [27]. The XML file, compatible with ADTool and OSEAD, containing the entire ADTree for tampering with the power meter is available at https://people.irisa.fr/Wojciech.Widel/studies/meter_study.zip.

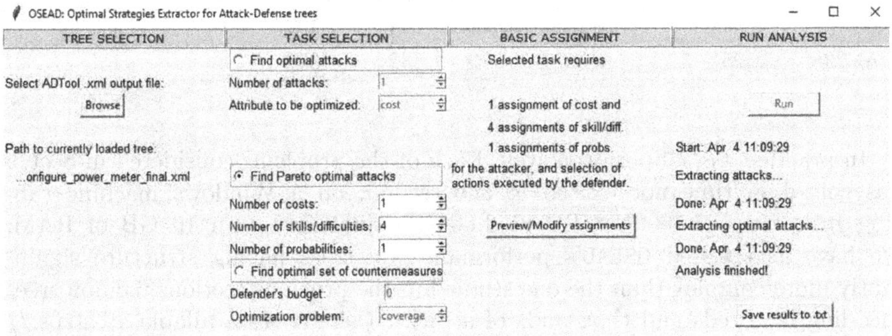

Fig. 9. OSEAD's main user interface

Once the tree is provided, users select the problem of interest, which is extraction of attacks that optimize a single parameter (tab *Find optimal attacks* in Fig. 9), attacks that are Pareto optimal (tab *Find Pareto optimal attacks*), or an optimal strategy of the defender (tab *Find optimal set of countermeasures*). The last step preceding the actual analysis is the assignment of values of parameters of interest to the basic actions in the tree. The values can be entered manually, imported from an XML file generated by ADTool, or loaded from a TXT file produced by OSEAD, as visualized in Fig. 10. With all the inputs provided, OSEAD solves the optimization problem specified by the user. The results obtained can be exported to a TXT file.

5.2 OSEAD's Performance

To solve the optimization problems, OSEAD first extracts attacks, as well as defender's strategies in the case of optimal set of countermeasures selection, from ADTree. Theoretically, this means that either a *set semantics* [8] or a *defense semantics* [24] of the tree is computed. The worst case size of both of these semantics is exponential in the number of basic actions of the tree. Another possible bottleneck in the process of determining an optimal strategy for the defender is solving the integer linear programming problem (ILP). OSEAD employs free ILP solver lp_solve [7] to achieve this task.

Table 7. OSEAD's runtime for determining Pareto optimal attacks

Name of file storing tree structure	Number of basic actions	Name of file storing basic assignment	Number of attacks	Number of Pareto optimal attacks	Runtime in seconds
tree03	16	*tree03_1_cost*	640	2	1
tree10	26	*tree10_1_cost*	14336	3	438
tree12	17	*tree12_5_costs*	2436	63	11
tree29	22	*tree29_5_costs*	640	304	1
tree30	23	*tree30_5_costs*	704	184	1
tree32	25	*tree32_5_costs*	832	378	1

In practice, OSEAD performs well. Each of the problems considered in Sect. 4 was solved in time not exceeding one second, on a Windows machine running Intel Core i7-5600U CPU at 2.60 GHz dual core with 16 GB of RAM. We have also tested OSEAD's performance on trees having structure significantly more complex than the one studied in the previous sections, i.e., on trees encoding hundreds and thousands of attacks. Using trees available at https://github.com/wwidel/pareto-tests/tree/master/trees and basic assignments given at https://github.com/wwidel/pareto-tests/tree/master/assignments, we have measured the time OSEAD needs to determine Pareto optimal attacks. An excerpt from the tests' results is presented in Table 7.

5.3 Implementation Details

OSEAD's computation engine and its user interface have been implemented in Python. Its architecture is depicted in Fig. 11. The implementation model con-

OSEAD: Optimal Strategies Extractor for Attack-Defense trees

Import   Export   Save & Exit

ATTACKER	cost_1 load from .xml	skill/diff_1 load from .xml	skill/diff_2 load from .xml	skill/diff_3 load from .xml	skill/diff_4 load from .xml	prob_1 load from .xml
acquire information from dumpster diving	0.0	0.0	0.0	0.0	1000.0	0.2
acquire information from public Internet source	0.0	0.0	1.0	0.0	100.0	0.79
bribe technician to reconfigure the power meter	500.0	0.0	0.0	3.0	10.0	0.52
bribe technician to reveal power meter credentials	300.0	0.0	0.0	2.0	10.0	0.5
buy optical probe	71.2	0.0	1.0	0.0	100.0	1.0
coerce technician into reconfiguring the power meter	0.0	0.0	0.0	3.0	100.0	0.3
coerce technician into revealing power meter credentials	0.0	0.0	0.0	3.0	100.0	0.33
collect information by exchanging gossips with employees	0.0	0.0	0.0	1.0	1000.0	0.46
enter power meter credentials	0.0	0.0	0.0	0.0	0.0	0.99
extract credentials	0.0	0.0	1.0	0.0	10.0	0.56
find and download software for hacking power meters	0.0	1.0	1.0	0.0	10.0	0.9
get employed as field technician	0.0	0.0	2.0	1.0	1000.0	0.48
get employed as intern by the energy provider	0.0	0.0	1.0	1.0	1000.0	0.52
have physical access to the power meter	0.0	0.0	0.0	0.0	0.0	1.0
intercept credentials	0.0	2.0	1.0	0.0	0.0	0.62
locate encrypted credentials in the dump	0.0	2.0	2.0	0.0	100.0	0.6
make optical probe	14.0	0.0	2.0	0.0	100.0	0.41
make the data dump from hardware component	0.0	1.0	3.0	0.0	100.0	0.73
monitor communication between hardware components	0.0	1.0	2.0	0.0	100.0	0.5
perform brute force attack	0.0	1.0	2.0	0.0	100.0	0.65
reconfigure power meter using authorized software/tools	0.0	0.0	1.0	0.0	10.0	0.94
reconfigure power meter using unauthorized software	0.0	0.0	2.0	0.0	100.0	0.75
select technician for obtaining power meter credentials	0.0	0.0	0.0	0.0	100.0	1.0
select technician for reconfiguring power meter	0.0	0.0	0.0	0.0	100.0	1.0
technician reconfigures power meter using authorized software/tools	0.0	0.0	0.0	0.0	10.0	1.0
trick technician into revealing power meter credentials	0.0	0.0	0.0	2.0	10.0	0.24
use optical probe to establish connection to the meter via the optical port	0.0	0.0	1.0	0.0	10.0	0.95

DEFENDER	performed
enforce policy of using strong passwords	☐
enforce policy to minimize Internet disclosure	☑
enforce policy to minimize leakage of physical artefacts	☐
limit the allowed number of invalid authentication attempts	☐
password authentication for establishing connection	☐
require authentication for introducing changes in power consumption configuration	☑
thorough background check before hiring new employees	☑
track popular social engineering tricks and warn personnel	☐

Fig. 10. Input management in OSEAD

sists of the *Tree Model* (storing the tree structure), the *Attribute Domain* (defining operations to be used when determining optimal attacks, e.g., (min, +) in the case of cost), the *ILP Problem* (derived from the *Tree Model*, using *defense semantics*, and storing the matrix of the selected optimization problem) and the *Basic Assignment* (storing values of parameters assigned to the basic actions).

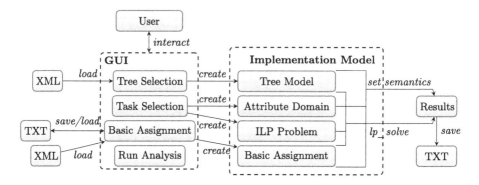

Fig. 11. An overview of the OSEAD architecture

OSEAD is open source and it runs on all main platforms. The version for Windows can be downloaded from https://people.irisa.fr/Wojciech.Widel/software/osead.zip. Using OSEAD on other platforms requires installing the *adtrees* Python package [36].

6   Conclusion

In this paper, we analyzed a real-life scenario of tampering with a power meter, using an ADTree. In addition to the actual evaluation of the power meters' security, this study allowed us to validate the quantitative analysis methods that we have recently developed for ADTrees with repeated labels [13,24,25]. To facilitate and automate their usage, we have implemented the OSEAD tool described in Sect. 5.

We took great care so that our model and analysis are as unbiased and impartial as possible. The tree was created by crossing several industrial and academic sources, and the input values estimation was performed by independent participants with various cultural background, from Estonia, France, Poland, and Russia.

As discussed in Sect. 4.3, we were able to confirm the intuitive conjecture about the practical importance of the multi-parameter analysis. We note that, despite the fact that the algorithms implemented in OSEAD are highly complex, the tool performs extremely well when applied to trees encoding hundreds of attacks, and reasonably well in the case of trees with up to several thousands of attacks.

This study corroborates practical usefulness of ADTrees in security and risk analysis. However, solutions for some pragmatic issues still need to be found. The bottleneck of our study was the attribution of parameter values to basic actions. While for some parameters, e.g., *cost*, finding an accurate estimate is easy (nowadays, it suffices to search on the Internet), for some other, e.g., *success probability*, this task is much more difficult, if not impossible. More research and practical investigation is definitely necessary before a reliable methodology for the estimation of values for basic actions can be proposed. In parallel, one could also investigate how sensitive the quantitative analysis methods proposed in [13, 24, 25] are wrt input values, i.e., how do errors propagate through the computations, if the input values used are not exact. In the light of its very promising efficiency, it seems that OSEAD will greatly facilitate performing this task, and this is the issue on which we will concentrate in the nearest future.

Finally, we would like to emphasize that an attack tree-based analysis, as the one performed in this case study, does not fully cover the entire process of risk analysis. For instance, a practical issue regarding Antoine's return on investment was not discussed in our work. This issue includes the analysis of the actual gain of Antoine versus the necessary expenses related to making the tempering possible, or the estimation of minimal time after which Antoine's investment in attacking the system would start to pay back. Also, one should not forget about a completely separate dimension of risk of being arrested for performing illegal tampering. Although we judged these aspects out of scope of our study, in real life they should be investigated before a truly optimal attack can be identified.

Acknowledgement. We would like to thank the following students and researchers for their (far from being trivial) contribution to the estimation of parameter values used in this study: Jean-Loup Hatchikian-Houdot (INSA Rennes, France), Pille Pullonen (Cybernetica AS, Estonia), Artur Riazanov (Saint Petersburg Department of V.A. Steklov Institute of Mathematics of the Russian Academy of Sciences, Russia), Petr Smirnov (Saint Petersburg State University, Russia), and Aivo Toots (Cybernetica AS, Estonia).

A   Optical Power Meter

Figure 12 shows a power meter with an optical port. Figure 13 illustrates how to connect to a power meter using an optical probe.

Fig. 12. A power meter with an optical port   (source: https://nl.wikipedia.org/wiki/IEC_62056)

Fig. 13. Optical probe connected to the power meter   (source: https://www.aliexpress.com/item/China-Manufacturer-DHL-free-Shipping-electricity-optical-meter-reading/32455842504.html?spm=2114.10010108.100009.1.6810cc24soIZC4&gps-id=pcDetailLeftTopSell&scm=1007.13482.95643.0&scm_id=1007.13482.95643.0&scm-url=1007.13482.95643.0&pvid=65873a85-f01b-4876-970d-b58b38041880)

References

1. ANSSI: La Méthode EBIOS Risk Manager (2018). https://www.ssi.gouv.fr/guide/la-methode-ebios-risk-manager-le-guide/
2. Arnold, F., Belinfante, A., Van der Berg, F., Guck, D., Stoelinga, M.: DFTCALC: a tool for efficient fault tree analysis. In: Bitsch, F., Guiochet, J., Kaâniche, M. (eds.) SAFECOMP 2013. LNCS, vol. 8153, pp. 293–301. Springer, Heidelberg (2013). https://doi.org/10.1007/978-3-642-40793-2_27
3. Arnold, F., Hermanns, H., Pulungan, R., Stoelinga, M.: Time-dependent analysis of attacks. In: Abadi, M., Kremer, S. (eds.) POST 2014. LNCS, vol. 8414, pp. 285–305. Springer, Heidelberg (2014). https://doi.org/10.1007/978-3-642-54792-8_16
4. Aslanyan, Z.: TREsPASS toolbox: attack tree evaluator (2016). Presentation of a tool developed for the EU project TREsPASS. https://vimeo.com/145070436. Accessed 16 Aug 2019

5. Aslanyan, Z., Nielson, F.: Pareto efficient solutions of attack-defence trees. In: Focardi, R., Myers, A. (eds.) POST 2015. LNCS, vol. 9036, pp. 95–114. Springer, Heidelberg (2015). https://doi.org/10.1007/978-3-662-46666-7_6
6. Bagnato, A., Kordy, B., Meland, P.H., Schweitzer, P.: Attribute decoration of attack–defense trees. IJSSE **3**(2), 1–35 (2012)
7. Berkelaar, M., Eikland, K., Notebaert, P.: lp_solve: Open source (Mixed-Integer) Linear Programming system (2005). Version 5.5.2.5, dated 24 September 2016. http://lpsolve.sourceforge.net/5.5/. Accessed 04 Apr 2019
8. Bossuat, A., Kordy, B.: Evil twins: handling repetitions in attack–defense trees. In: Liu, P., Mauw, S., Stølen, K. (eds.) GraMSec 2017. LNCS, vol. 10744, pp. 17–37. Springer, Cham (2018). https://doi.org/10.1007/978-3-319-74860-3_2
9. Carpenter, M.: Advanced metering infrastructure attack methodology (2009). http://docshare.tips/ami-attack-methodology_5849023fb6d87fd2bb8b4806.html. Accessed 20 Feb 2019
10. Dürrwang, J., Braun, J., Rumez, M., Kriesten, R., Pretschner, A.: Enhancement of automotive penetration testing with threat analyses results. SAE Int. J. Cybersecurity **1**, 91–112 (2018). https://doi.org/10.4271/11-01-02-0005
11. EAC Advisory Board and Standards Board: Election Operations Assessment - Threat Trees and Matrices and Threat Instance Risk Analyzer (TIRA) (2009). https://www.eac.gov/assets/1/28/Election_Operations_Assessment_Threat_Trees_and_Matrices_and_Threat_Instance_Risk_Analyzer_(TIRA).pdf. Accessed 13 June 2018
12. Edge, K.S., Dalton II, G.C., Raines, R.A., Mills, R.F.: Using attack and protection trees to analyze threats and defenses to homeland security. In: MILCOM, pp. 1–7. IEEE (2006)
13. Fila, B., Wideł, W.: Efficient attack–defense tree analysis using Pareto attribute domains. In: CSF, pp. 200–215. IEEE Computer Society (2019)
14. Fraile, M., Ford, M., Gadyatskaya, O., Kumar, R., Stoelinga, M., Trujillo-Rasua, R.: Using attack-defense trees to analyze threats and countermeasures in an ATM: a case study. In: Horkoff, J., Jeusfeld, M.A., Persson, A. (eds.) PoEM 2016. LNBIP, vol. 267, pp. 326–334. Springer, Cham (2016). https://doi.org/10.1007/978-3-319-48393-1_24
15. Frederic Byumvuhore: FEATURED: REG steps up crackdown on electricity theft (2019). https://www.newtimes.co.rw/news/featured-reg-steps-crackdown-electricity-theft. Accessed 05 Apr 2019
16. Gadyatskaya, O., Hansen, R.R., Larsen, K.G., Legay, A., Olesen, M.C., Poulsen, D.B.: Modelling attack-defense trees using timed automata. In: Fränzle, M., Markey, N. (eds.) FORMATS 2016. LNCS, vol. 9884, pp. 35–50. Springer, Cham (2016). https://doi.org/10.1007/978-3-319-44878-7_3
17. Gadyatskaya, O., Jhawar, R., Kordy, P., Lounis, K., Mauw, S., Trujillo-Rasua, R.: Attack trees for practical security assessment: ranking of attack scenarios with ADTool 2.0. In: Agha, G., Van Houdt, B. (eds.) QEST 2016. LNCS, vol. 9826, pp. 159–162. Springer, Cham (2016). https://doi.org/10.1007/978-3-319-43425-4_10
18. hashcat (2016). https://hashcat.net/hashcat/. Accessed 27 Mar 2019
19. Kelly-Detwiler, P.: Electricity theft: a bigger issue than you think (2013). https://www.forbes.com/sites/peterdetwiler/2013/04/23/electricity-theft-a-bigger-issue-than-you-think/#5475872972ef. Accessed 20 Feb 2019
20. Wilburg, K.: GPL lost US$450M in 19 years to electricity theft, poor networks (2018). https://www.kaieteurnewsonline.com/2018/12/10/gpl-lost-us450m-in-19-years-to-electricity-theft-poor-networks/. Accessed 05 Apr 2019

21. Kordy, B., Mauw, S., Radomirovic, S., Schweitzer, P.: Attack–defense trees. J. Log. Comput. **24**(1), 55–87 (2014)
22. Kordy, B., Mauw, S., Schweitzer, P.: Quantitative questions on attack–defense trees. In: Kwon, T., Lee, M.-K., Kwon, D. (eds.) ICISC 2012. LNCS, vol. 7839, pp. 49–64. Springer, Heidelberg (2013). https://doi.org/10.1007/978-3-642-37682-5_5
23. Kordy, B., Pouly, M., Schweitzer, P.: Probabilistic reasoning with graphical security models. Inf. Sci. **342**, 111–131 (2016)
24. Kordy, B., Wideł, W.: How well can I secure my system? In: Polikarpova, N., Schneider, S. (eds.) IFM 2017. LNCS, vol. 10510, pp. 332–347. Springer, Cham (2017). https://doi.org/10.1007/978-3-319-66845-1_22
25. Kordy, B., Wideł, W.: On quantitative analysis of attack–defense trees with repeated labels. In: Bauer, L., Küsters, R. (eds.) POST 2018. LNCS, vol. 10804, pp. 325–346. Springer, Cham (2018). https://doi.org/10.1007/978-3-319-89722-6_14
26. Krebs, B.: FBI: Smart Meter Hacks Likely to Spread (2012). https://krebsonsecurity.com/2012/04/fbi-smart-meter-hacks-likely-to-spread/. Accessed 20 Feb 2019
27. Kumar, R., et al.: Effective analysis of attack trees: a model-driven approach. In: Russo, A., Schürr, A. (eds.) FASE 2018. LNCS, vol. 10802, pp. 56–73. Springer, Cham (2018). https://doi.org/10.1007/978-3-319-89363-1_4
28. LLC, N.G.: World Loses $89.3 Billion to Electricity Theft Annually, $58.7 Billion in Emerging Markets (2014). https://www.prnewswire.com/news-releases/world-loses-893-billion-to-electricity-theft-annually-587-billion-in-emerging-markets-300006515.html. Accessed 20 Feb 2019
29. Lund, M.S., Solhaug, B., Stølen, K.: Model-Driven Risk Analysis: The CORAS Approach. Springer, Heidelberg (2011). https://doi.org/10.1007/978-3-642-12323-8
30. Mauw, S., Oostdijk, M.: Foundations of attack trees. In: Won, D.H., Kim, S. (eds.) ICISC 2005. LNCS, vol. 3935, pp. 186–198. Springer, Heidelberg (2006). https://doi.org/10.1007/11734727_17
31. McCullough, J.: Deterrent and detection of smart grid meter tampering and theft of electricity, water, or gas (2010). https://www.elstersolutions.com/assets/downloads/WP42-1010A.pdf. Accessed 20 Feb 2019
32. Ms. Smith: FBI Warns Smart Meter Hacking May Cost Utility Companies $400 Million A Year (2012). https://www.csoonline.com/article/2222111/fbi-warns-smart-meter-hacking-may-cost-utility-companies–400-million-a-year.html. Accessed 05 Apr 2019
33. National Electric Sector Cybersecurity Organization Resource (NESCOR): Analysis of selected electric sector high risk failure scenarios, version 2.0 (2015). http://smartgrid.epri.com/doc/NESCOR%20Detailed%20Failure%20Scenarios%20v2.pdf
34. Ophcrack (2016). http://ophcrack.sourceforge.net/. Accessed 17 Mar 2017
35. Ophcrack (2016). https://www.openwall.com/john/. Accessed 27 Mar 2019
36. adtrees Python package (2019). https://github.com/wwidel/adtrees. Accessed 05 Apr 2019
37. Refsdal, A., Solhaug, B., Stølen, K.: Cyber-Risk Management. Springer, Cham (2015). https://doi.org/10.1007/978-3-319-23570-7
38. Schneier, B.: Attack trees: modeling security threats. Dr. Dobb's J. Softw. Tools **24**(12), 21–29 (1999)
39. T&D World: India To Spend $21.6 Billion On Smart Grid Infrastructure By 2025 (2015). https://www.tdworld.com/smart-grid/india-spend-216-billion-smart-grid-infrastructure-2025. Accessed 05 Apr 2019

40. Weber, D.C.: Optiguard: A Smart Meter Assessment Toolkit (2012). https://media.blackhat.com/bh-us-12/Briefings/Weber/BH_US_12_Weber_Eye_of_the_Meter_WP.pdf. Accessed 20 Feb 2019

41. Wideł, W., Audinot, M., Fila, B., Pinchinat, S.: Beyond 2014: formal methods for attack tree-based security modeling. ACM Comput. Surv. **52**(4), 75:1–75:36 (2019). https://doi.org/10.1145/3331524

Risk Management and Attack Graphs

Quantifying and Analyzing Information Security Risk from Incident Data

Gaute Wangen[(✉)]

Norwegian University of Science and Technology,
Teknologiveien 22, 2802 Gjøvik, Norway
gaute.wangen@ntnu.no

Abstract. Multiple cybersecurity risk assessment and root cause analysis methods propose incident data as a source of information. However, it is not a straightforward matter to apply incident data in risk assessments. The paper trail of incident data is often incomplete, ambiguous, and dependent on the incident handlers routines for keeping records. Current incident classification approaches classify incidents as one specific type, for example, "Data spillage," "Compromised information," or "Hacking." Through incident analysis, we found that the current classification schemes are ambiguous and that most incident consists of additional components. This paper builds on previous work on incident classifications and proposes a method for quantifying and risk analyzing incident data for improving decision-making. The applied approach uses a set of incident data to derive the causes, outcomes, and frequencies of risk events. The data in this paper was gathered from a year of incident handling from a Scandinavian university's security operations center (SOC), and consists of 550 handled incidents from November 2016 to October 2017. By applying the proposed method, this paper offers empirical insight into the risk frequencies of the University during the period. We demonstrate the utility of the approach by deducting the properties of the most frequent risks and creating graphical representations of risks using a bow-tie diagram. The primary contribution of this paper is the highlighting of the ambiguity of existing incident classification methods and how to address it in risk quantification. Additionally, we apply the data in risk analysis to provide insight into common cyber risks faced by the University during the period. A fundamental limitation is that this study only defines adverse outcomes and does not include consequence estimates.

Keywords: Information security · Cyber security · Security incidents · Risk analysis · Threat intelligence

1 Introduction

The topic of this paper is how to categorize, quantify, and apply an organization's information security (InfoSec) incident register for risk analysis. An InfoSec incident is an event or occurrence that contains a breach of either confidentiality,

© Springer Nature Switzerland AG 2019
M. Albanese et al. (Eds.): GraMSec 2019, LNCS 11720, pp. 129–154, 2019.
https://doi.org/10.1007/978-3-030-36537-0_7

integrity, or availability of an information asset or a service. A computer security incident response team (CSRIT) or Security Operations Centre (SOC) incident typically handles by the incident. Such teams usually consist of InfoSec experts who aim to resolve the event and return the system to normal operations. The SOC often uses incident management systems to maintain records of the incident and steps taken to resolve it. A generic example of an incident handling process is as follows: an incident gets reported to or gets detected by the SOC and assigned to an incident handler (IH). The IH determines if it is an incident and if it belongs to the SOC. If so, he attempts to resolve the incident and records actions taken. This process makes incident data a readily available data source in many organizations for improving the InfoSec risk analysis (ISRA). The risk analysis conducted in this study consists of identifying and quantifying issues and analyzing their significance. Quantification is the act of counting and measuring observations into quantities. Previous work has gone into the analysis of InfoSec incidents and their cost [6,7,12,14], but the literature is scarce on the ambiguity of incident data and the process of how to adopt them into ISRA. The following sequence of events describes the core of the problem: "A phishing e-mail arrives in an employee inbox. The e-mail contains a malicious attachment. The employee opens the attachment, and a malware trojan infects the computer and connects back to the attacker. The attacker uploads a keylogger to the infected computer to extract the company username and password. The attacker later uses the stolen credentials to log into the company network to look for vulnerable servers from which he can exfiltrate information." The industry practice is to classify an incident under one specific category as proposed by FIRST[1], US-CERT[2], and others [1,3,5,6,12]. For example, FIRST proposes Denial of service, Forensics, Compromised Information, Compromised Asset, Unlawful activity, Internal/External hacking, Malware, E-mail, Consulting, and Policy Violations as categories. However, how does one classify the described incident? Is it an e-mail attack, malware infection, compromised information, hacking, or compromised asset? It is arguably all of them. The inherent ambiguity of incident classifications is not sufficiently addressed in current research. There is more useful data that can be used for knowledge gathering and risk analysis in the incident registers.

There are additional obstacles to quantifying incident data which we address in this study: Firstly, no two incidents are identical, and it is rare to find a data record with a complete incident scenario as previously described. To be able to quantify the incidents, we first need to categorize them in a meaningful manner. There already exists classification frameworks, such as those proposed by FIRST and CERT-US, together with taxonomies of computer security incidents (e.g. [8,11]) that provide a starting point. For this study, we operate with two levels of granularity for the classifications. As previously discussed, a computer

---

[1] Forum of Incident Response and Security Teams https://www.first.org/resources/guides/csirt_case_classification.html (Visited May 2019).

[2] Federal Incident Reporting Guidelines https://www.us-cert.gov/government-users/reporting-requirements (Visited May 2019).

incident consists of more information that can be quantified for decision-making than can be described by a one incident classification (e.g., "Data leakage"). The previously mentioned classification schemes do not sufficiently recognize the usefulness of adapting more parts of the incident data into risk analysis. Although a typical security incident consists of more than two steps, our data analysis revealed that for the majority of cases, we could deduce both a cause and an outcome. Where the former relates to the threat, attack vector and vulnerability, and the latter relates to the malicious action taken, intent, and outcome. In terms of security controls, the former relates to preventive barriers and the latter relates to reactive barriers.

Through a study conducted at the SOC in the Norwegian University of Science and Technology (NTNU) we have gathered data from 550 incidents. The main purpose of this paper is to show how the risk analyst can categorize and quantify incident data into cause and outcomes, and apply in risk analysis. Specifically, we address the following research questions:

1. How can incident data be quantified and applied in risk analysis?
2. What does the risk picture look like at the University based on the incident data?
3. How can incident data be applied to graphical risk analysis?

We address the research question (i) by proposing a risk classification and quantification scheme. Research question (ii) is addressed by analyzing the quantified data from the case study. The final research question (iii) is answered by applying the quantified incident data in different risk analysis schemes to demonstrate the utility and extract knowledge about a specific set of risks.

The remainder of the paper is organized as follows, Sect. 2 provides general background information on incident classification and quantitative risk assessment. In Sect. 3, we describe how the framework was developed and applied. Furthermore, we describe the data collection process and the applied classifications and risk analysis. Section 4 presents the study with the risk picture for the institution from applying the proposed method. Furthermore, this paper extends the risk analysis in Sect. 5 where specific risks are studied as examples. Section 6 evaluates the proposed method including the limitations of the study and the proposals for future work. Lastly, we conclude the work in Sect. 7.

2   Background and Related Work

Firstly, this section presents the preliminary work and relevant reports applied in this study. Furthermore, we address the previous work on existing InfoSec incident classification. Lastly, we survey the relevant literature on risk quantification for InfoSec.

This paper builds on the preliminary work in incident classification for root cause analysis published by Hellesen et al. [9] and use of the critical incident tool. Additionally, Chapman [5] has published a policy notice on cyber-security in higher education where he outlines key security challenges faced by the UK higher education and research. Chapman also includes incident statistics from the UK based Janet network in the period of January - December 2018. This paper also applies results from the technical report "Unrecorded security incidents at NTNU 2018" (Norwegian report) [16] which contains statistics from a security awareness survey conducted at the University (N = 532, Margin of error = 4% at 95% confidence interval). We use both the statistics from the Janet network and the technical report for comparison in this paper.

Kjaerland proposes a taxonomy of computer security incidents based on *Method of operation* and *Impact* which recognizes the attack and the impact as components of the incident [11]. The Method of operation category consists of malicious actions and attack vectors an attacker can apply. The Impact variable consists of the attacker's intentions, such as disrupt and destruct. Kjaerland's approach provides a starting point for incident classifications. Hansman and Hunt [8] proposes a technical taxonomy for incident classification, containing four dimensions or six levels of categorization per incident. The information gathered for this taxonomy is useful for deep analysis of each incident, but the proposed level of technical detail is typically not needed for risk quantification. ENISA's *Reference Incident Classification Taxonomy* [3] provides the basic categories for the proposed framework. While the *The Common Taxonomy for Law Enforcement and The National Network of CSIRTs* [1] was published by ENISA and Interpol to bridge the gap between the CSIRTs and international Law Enforcement communities. It adds a legislative framework to facilitate the harmonization of incident reporting to competent authorities, the development of useful statistics and sharing information within the entire cybercrime ecosystem. It proposes nine broad categories for incident classification with sub-classifications based on malicious actions. Common for all of these approaches is that they are not developed specifically for risk classification. However, they provide a solid starting point for an incident classification framework scoped for risk assessment.

One of the primary goals of this paper is to quantify the frequencies of occurrences for InfoSec risks. Recently there have been multiple attempts at quantifying the cost of information risk incidents. The trend in loss quantification has been for security vendors and other parties responsible for surveys to publish loss estimates. Florencio and Herley [7] discuss weaknesses such as cyber-crime surveys and how the results can entail large amounts of uncertainty. Edwards et al. [6] investigate a similar problem confined to reported data record breaches from 2006 to 2015. The authors focus on the likelihood component and demonstrate how to derive estimates and predictions about data breaches. The analyzed breaches in the study are divided into negligent and malicious breaches with eight sub-categories. Similar to the study in this paper, Kuypers et al. [12] tackles the problem of incident classification and analysis using eight categories for incident classification. Kuypers also divides the incidents into severity based on the time spent on handling each incident. The study utilizes 60,000 incident records collected over six years. The scope of the Kuypers et al. paper is primarily quantification for predicting trends of frequencies of events and losses. This scope differs from the current paper in that we conduct deeper analysis by investigating the threat action and applying typical InfoSec risk analysis methods [17] to the data.

Bernsmed et al. [4] have published an illustrative paper for applying bow-tie analysis in cybersecurity. Although, a slightly different approach than Bernsmed, we found the best-suited risk approach for the incident data was the *Bow-tie* as it utilizes both causes and outcomes of each risk or incident. Furthermore, this study builds on Wangen et al.'s [17] Core Unified Risk Framework (CURF), which is a bottom-up classification of ISRA methods and contains an extensive overview of ISRA method content.

3   Method

This section outlines the applied method for the study, starting with classification framework development. Furthermore, this section describes data collection, risk analysis and statistics.

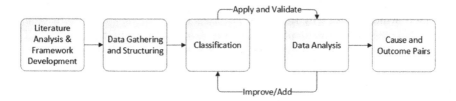

Fig. 1. Development scheme for Incident classification framework

3.1   The Incident Classification and Analysis Scheme

As mentioned in the related work, there exist multiple frameworks for classifying computer security incidents, and the proposed classification scheme builds on these. There are some underlying premises to the framework: An incident must have (at least) one cause and one outcome. Where the cause relates to a threat exploiting a vulnerability. The outcome is the empirically observable malicious action taken by the threat where he acts on objectives. This premise is a simplification of Lockheed Martin's seven-step cyber kill chain.[3] No two incidents are identical, which means that we must classify to quantify. Specifically, the framework idea is as follows: Start with a set of level 1 incident categories tailored to the organization. Furthermore, the risk analyst identifies a cause and an outcome for each reviewed incident and quantifies them. If either the identified cause or outcome is not present in the current set, the analyst either adds it to the classification set or embeds it into an existing category. By applying this approach, the

---

[3] Lockheed Martin, *Cyber Kill Chain* https://www.lockheedmartin.com/en-us/capabilities/cyber/cyber-kill-chain.html (Visited May 2019).

classification framework receives continuous validation and improvement, visualized in Fig. 1. For practical reasons, we have chosen to define both the cause and outcome within the same categories. The classification scheme applied in this study consists of fourteen main classifications with sub-classifications. For simple incident classification, it might be sufficient to apply the level 1 categories (e.g. [6,12]). For the dataset in this study, the applied incident classifications are listed in Table 2, which shows that some incident categories are primarily applied for cause classifications. While other categories are primarily outcome related, and some are overlapping.

For the data collected in this study, the paper trail of an incident consists of the original incident trigger, the IH's log of steps taken to resolve the incident, and all correspondence with affected parties. There might be uncertainty regarding the initial cause in some of the incidents. For these cases, we solved this problem by adding broader categories, such as "System compromise," to quantify the incidents where we have limited knowledge. A limitation of the incident data is that the record sometimes does not include the real cause. For example, an incident cause might be a trojan virus and where the observable outcome is data exfiltration. However, the cause of the initial infection may be unknown, which is also why that in some cases, for example, a compromised user can be the cause of an incident and in other cases it might be the outcome. The level 1 categories we have applied for classification are described in Table 1.

3.2   Data Collection

The NTNU SOC was established in late 2016 and was building capabilities during the time of data collection. All the data presented in this study was extracted from the NTNU SOC's incident management system, "Request Tracker for Incident Response" (RTIR) from Best Practical Solutions. The dataset presented in this study includes all incidents handled by the SOC between Nov 2016 and Oct 2017, counting 550 incidents in total. The dataset includes incidents triggered by in-house capabilities, user reported, and third-party reports. The latter include, for example, other CSIRT notifications, users, and vulnerability reports. At the time of data collection, the NTNU SOC consisted of 4 dedicated incident handlers and one part-time member dedicated to working with email-related issues. We qualitatively analyzed each of the incident reports and assigned a cause and an outcome within the classification scheme described previously.

3.3   Risk Analysis and Statistics

The ISRA approach in this paper primarily builds on the ISO/IEC 27005:2011 (ISO27005) [2] for understanding risk management. A risk in our proposed ISRA consists of a scenario with an adverse outcome, with a probability distribution of consequences. The ISO27005 defines the scenario as a combination of assets, vulnerability, threat, controls, and outcome. The analysis in this paper quantifies causes and outcomes and determines the frequency of occurrence for each separately, and cause and outcome pairs together. The study applied IBM SPSS

Table 1. Incident classification descriptions applied in the study.

No	Level 1	Description
1	Abuse	Abuse refers to the improper or wrongful use of company assets. Which includes spamming using company resources. It can also be hosting illegal content on the company network, misusing access rights granted, or users complaining about abuse
2	Unlawful activity	Refers to any activity that is deemed illegal either by law and legislation. This category also includes police petitions on data extradition
3	Malware	The malware category is broad and contains multiple sub-classifications of different malware categories. As there are many strains of malware, the sub-classification is limited to address our incident and risk analysis needs
4	Reconnaissance	The reconnaissance category relates to incidents triggered by typical adversarial information-gathering activities, including network scanning and packet sniffing
5	Compromised Asset	A compromised asset refers to a company asset that has been breached and is under adversarial control. The classification scheme applies to four sub-categories. An asset or system in this setting refers primarily to servers, computers, smartphones, and tablets. Network device refers to network infrastructures, such as routers, printers, raspberry pies, and other networked devices. While application compromise refers to the breach of a specific application. This category also includes hardware theft
6	Compromised User	A compromised user is when the username and password of an account get compromised. The level 2 category separates between admin and regular users based on the difference in consequence, where the former constitutes a more severe breach
7	Compromised Information	This category is used for incidents triggered by observed adversarial actions, such as leaking sensitive data, modifying or changing information, unauthorized access, and privacy violations
8	Vulnerable Asset	A vulnerable asset is an organizational property that is vulnerable to external and internal attacks. Typically, software or system can be vulnerable due to a new vulnerability or lack of patch management. Alternatively, there can be a misconfiguration that leaves the asset vulnerable, or it can also be an inherent vulnerability in a protocol or similar that leaves the asset open for abuse
9	Denial of Service	Denial of service (DoS) occurs when a service or asset becomes unavailable. A DoS can be distributed (DDOS) from many compromised systems to route traffic to the target or can occur just from a single machine. A DoS can be caused intentionally by an attacker, or there might be an outage or another failure that causes it. One of the tougher issues to tackle in the classification is whether participation in an outgoing DDoS from a vulnerable asset (e.g., reflexive attack)

(continued)

Table 1. (*continued*)

No	Level 1	Description
10	Social Engineering	Typical social engineering attacks are phishing and spear-phishing. Where the former targets organization-wide and the latter target specific individuals or groups. Whaling and CEO frauds target CEOs, high ranking company officers, and their co-workers. The category also includes less frequent frauds such as support fraud, phone fraud, and SMS fraud
11	Intrusion Attempt	Intrusion attempts are when the adversary attempts to penetrate the system using technical or physical means. Typical examples are brute-force attempts on login screens and executing exploits on seemingly vulnerable systems. Other incident triggers can come from sensor alarms (intrusion detection systems), log analysis, honeypots, and other detection technology
12	Policy Violation	For the NTNU case data, we have two overarching policies: Information Security Policy and the IT Policy. Each with his management system consisting of standards, rules, and procedures to follow. Violations of any of them can trigger an incident
13	Other	The other category is for security incidents that do not classify in any of the above but still needs to be solved by the SOC
14	Outcome: Negligible/ Fixed/ Failed Attack	This category is only for classifying outcomes and is necessary for the SOC solved incidents that either has no observable adverse outcome or qualified as a failed attack. E.g., a spam campaign launched by an external attacker that quickly gets blocked and handled by the SOC without any employees receiving or opening links

for descriptive statistics. The causes and outcomes are analyzed using descriptive statistics, histograms, time-series, and cause-effect flow charts. We apply confidence intervals (CI) for examples of event prediction where we apply the mean and 90% CI as proposed for risk analysis by Seiersen and Hubbard [10]. As proposed in the preliminary study by Hellesen et al. [9] the incident data has utility in obtaining knowledge about causes of unwanted outcomes and vice verse. We apply bow-tie diagrams to illustrate the utility of the data set. Bow-tie is a visual representation of the "attack flow," illustrating the causes, preventive controls, reactive controls, and outcomes, see Fig. 7. The bow-tie diagrams allow for utilization of both cause and outcome for each incident, thus enabling more in-depth analysis of each risk. The bow-tie analysis is a representation of the attack flow starting on the left with an attack. The diagram is then used to illustrate the security controls in place to prevent the unwanted incident, which occurs if all the preventive controls fail. Furthermore, the mitigating controls in place to reduce the consequence are listed. Again, if these fail, an unwanted outcome will occur, which constitute an incident.

To further illustrate the utility of the incident data in traditional ISRA, we apply the ISO27005 [2] approach which builds on the asset, threat, and vulnerability scheme for security risks. A lot of the focus in the InfoSec industry is on the threat [17,18], so, we explicitly show how to use the incident data for the threat assessment. For this analysis, the premise is that the cause relates to the threat, the attack method, and vulnerability. The incident data is a historical record which allows us to work with risk analysis: Starting with formulating the risks from cause and outcome pairs, and then decomposing the data into assets, threats and vulnerabilities by examining which assets were compromised, exploited vulnerabilities, and examining the malicious actions taken by the attacker once inside the system. The outcome reveals either the motive, targeted asset, threat actor class and intent. The threat actor is broadly classified, and the risk scenario is defined as proposed by Potter's *Practical Threat modeling* [13]. Potter emphasizes that a threat is as specific as it needs to be; for example, in most cases of threat modeling, it does not make sense to divide threat actors into groups of high granularity. If we are mainly working on defense grouping on motivation as proposed by Potter is generally sufficient for deriving security requirements: "ACTOR does ACTION to ASSET for OUTCOME because of MOTIVATION." The threat actor categories proposed by Potter are *Nation state (APT)*, *Organized Crime*, *Insiders*, *Hacktivists*, *Script Kiddies*, and *Others*.

4   The InfoSec Risk Picture at the University

The case data, together with relevant available statistics, were collected from the SOC at the Norwegian University of Science and Technology (NTNU). At the time of the study, the SOC constituency amounted to about 39 700 students and 6 900 full-time equivalent staff. There were approximately 1500 servers and 15000 managed clients in the network. NTNU has eight faculties and academic curriculum in the natural sciences, social sciences, teacher education, humanities, medicine and health sciences, economics, finance, and administration, as well as architecture and the arts. NTNU also provides state-of-the-art research within multiple technology fields which dictates confidentiality requirements.

The following sections outline the results of applying the classification scheme proposed in this paper, starting with the overarching distributions (level 1) of causes and outcomes. Further, we provide the distributions applying the level 2 categorizations with more granularity. Lastly, we derive the most frequent cause and outcome combinations. For each step, we give examples of how to extract useful information for decision-making from the data.

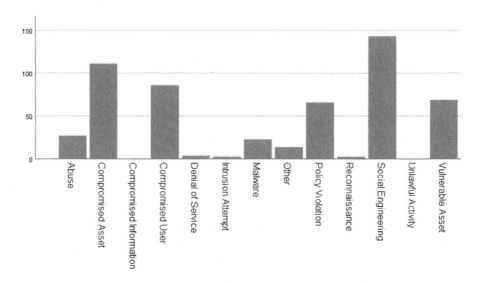

Fig. 2. Incident causes in the NTNU SOC

4.1 Risks According to the Incident Data

The following results start at the top level with the primary causes and outcomes of incidents in their separate histograms. We then describe the level 2 causes and outcomes in a table, before showing the level 1 connection between causes and outcomes. Lastly, we show the trends of causes and outcomes throughout the data collection period (Table 4).

The total distribution of the causes is illustrated in Fig. 2, which provides an overarching picture of the most common causes of incidents at NTNU. The most common causes in the data set are social engineering attempts (143), compromised assets (107), and compromised users (87). No incidents were caused by actions related to unlawful activity or detection of compromised information.

The total distribution of the 550 outcomes is illustrated in Fig. 2. 201 incidents were handled with a negligible outcome. Furthermore, Abuse (84), Denial of Service (68), and Unlawful activity (55) are the most frequent outcomes of incidents in the constituency.

Table 2 provides the level 2 distribution of both the causes and outcomes for the incidents. A thing to note is that the table only shows quantities and does not contain the connection between the numbers in each column. These numbers provide a higher granularity of information than the level 1 distributions, for example, by decomposing the abuse classification we reveal that the most frequent type of abuse is spam for both causes and outcomes, with misuse of company resources (28) as a common outcome of incidents. The majority of the misuse incidents were caused by a hacking campaign called the *Silent Librarian* by Chapman [5]. The Silent librarian campaign was launched against Universities and abused the access to publication channels to mass download research

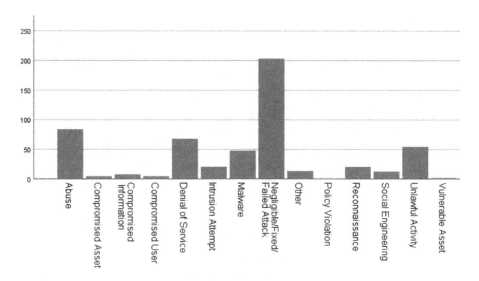

Fig. 3. Incident outcomes in the NTNU SOC

articles. In his article, Chapman attributes the campaign to the Iranian state-sponsored group *Mabna* hacking group. The System compromise category is a frequent cause for incidents (102), while it is seldom an observable outcome of an incident (3). Furthermore, we observe that unlawful activity, reconnaissance, and compromised information are primarily outcome-categories, while the vulnerable asset, compromised asset, social engineering, and policy violations are primarily cause-categories.

Table 3 reveals the level 1 connection between the causes and outcomes in the dataset. The cause is listed on the y-axis and outcome on the X-axis. The table lists the combinations of causes and outcomes for the dataset and describes the frequency of occurrence for each combination. For example, a compromised asset has multiple outcomes, whereas twelve incidents of compromise lead to abuse, thirty-three to malware infections, and sixteen to further reconnaissance. There are several common combinations in the dataset, such as social engineering attempts and fixed attacks (127), which is typically phishing attacks (Table 2). Vulnerable assets often relate to protocol vulnerabilities and similar weaknesses exploited in amplification attacks (29).

4.2   Trends and Predictions

Tables 4 and 5 illustrate the development of level 1 causes and outcomes throughout the data collection period. For example, the most frequent cause of incidents in Table 4 is social engineering attempts. These attacks occur regularly each month throughout the dataset with an uptick the three last months, $N = 143$ with min $= 4$ and max $= 23$. The quality of the phishing attacks are often attempts of tricking employees into giving away their passwords, visit malicious

Table 2. Frequencies of 550 incidents categorized on Cause and Outcome. No link between the Cause and Outcome columns.

Classification (Level 1)	Sub-Classification (Level 2)	Cause	Outcome
Abuse (1)	Spam	25	53
	Illegal Content		1
	User Complaint	1	2
	Misuse		28
	Web Site copying	1	
Unlawful activity (2)	Copyright/Piracy		55
Malware (3)	Virus	1	3
	Worm		1
	Backdoor/Rootkit		2
	Trojan	15	31
	Spyware/Adware	1	
	Hacking tools, Exploits, & Exploit kits	1	1
	Ransomware	3	6
	DNS Hijack		2
	Unspecified	2	2
Reconnaissance (4)	Scanning	3	21
Compromised Asset (5)	System Compromise	102	3
	Network Device Compromise	4	
	Application Compromise		1
	Hardware theft	1	1
Compromised User (6)	Admin User compromise	2	
	Regular User compromise	84	5
Compromised Information (7)	Data leakage		3
	Unauthorised Modification		2
	Unauthorised Access		1
	Privacy Violation		2
Vulnerable Asset (8)	Misconfiguration	14	
	Vulnerable Software	13	
	Vulnerable System	2	
	0-Day Vulnerability	1	
	Open for abuse	41	2

(continued)

Table 2. (*continued*)

Classification (Level 1)	Sub-Classification (Level 2)	Cause	Outcome
Denial of Service (9)	DoS/DDoS	2	7
	DoS/DDoS Outgoing	2	61
Social Engineering (10)	Phishing	112	10
	Spear Phishing	12	
	Whaling/CEO Fraud	19	2
Intrusion Attempt (11)	Brute Force	2	19
	Exploit on non-vulnerable system	1	2
Policy Violation (12)	Information Security Policy	6	
	IT Policy	60	1
Other (13)	Unclassified	16	16
	Hoax	1	
	Malware Hosting		1
Negligible/Fixed/Failed Attack (14)			203
Sum		550	550

Table 3. Cause (Y-axis) and Outcome (X-axis) combinations between Level 1 categories

Cause	Outcome													
	Abuse	Unlaw. Act.	Malware	Recon	Comp. Asset	Comp. User	Comp. Info.	Vuln. Asset	DoS	Social Engi.	Intr. Att.	Policy Viola.	Other	Fixed/ Failed
Abuse	0	0	0	0	0	1	0	0	0	0	0	0	1	25
Unlawful Act.	0	0	0	0	0	0	0	0	0	0	0	0	0	0
Malware	0	0	2	0	0	0	2	0	10	0	3	0	0	6
Recon.	0	0	0	0	0	0	0	0	1	0	0	0	0	2
Comp. Asset.	12	0	34	18	2	0	1	0	21	3	16	0	1	3
Comp. User	68	1	1	0	0	0	0	0	1	9	2	0	1	3
Comp. Info	0	0	0	0	0	0	0	0	0	0	0	0	0	0
Vuln. Asset	0	0	1	1	2	0	2	2	29	0	0	1	1	30
DoS	0	0	0	0	0	0	0	0	4	0	0	0	0	0
Soc. Eng.	2	0	9	0	0	4	0	0	0	0	0	0	1	127
Intr. Attempt	0	0	0	1	0	0	0	0	1	0	0	0	0	1
Policy Vio.	2	54	1	1	1	0	2	0	0	1	0	0	0	4
Other	0	0	0	0	0	0	1	0	0	0	0	0	9	2

webpages, or open malicious attachments. There is also evidence to suggest that the incidents in the dataset are only the tip of the iceberg and that this is an everyday event at the University [16]. The majority of these attacks originate externally and seem to motivated by financial gain.

The compromised asset class is the second most frequent cause of incidents, N = 111 with min = 3 and max = 19 values. When an asset gets compromised, the data shows that the most frequent course of action taken by the attacker is to install Trojan malware (29 occurrences). This action gets detected and handled by the SOC when the malware attempts to call home, so we do not know more about the intent in these cases. Other frequent outcomes is the compromised asset gets exploited in outgoing DDoS attacks (21 occurrences), or used as a stepping stone in either scanning (18) or brute force attacks (14).

Furthermore, the data shows that compromised users are the consistent cause of incidents throughout the year averaging seven incidents per month. A frequent course of action taken by the attacker is to abuse the compromised user account to distribute spam e-mail on the internal network (42 cases) implying a financial motivation. The data also reveals more malicious attempts of abusing the account for social engineering such as phishing (6 attempts) and whaling/CEO fraud (2 attempts).

Considering the results in the incident outcome Table 5, the data show an increase in abuse cases in the spring semester (Jan - Jun mean = 10 per month) compared to the autumn semester (Nov-Dec and Jul-Oct, mean = 4). Among other things, these abuse cases were very likely caused by the *Silent librarian* campaign, which consisted of a wave of attacks involving compromised user accounts and exploitation of the access given by the University to harvest research articles (26 accounts).

Table 4. Yearly development in incident causes

	NOV	DEC	JAN	FEB	MAR	APR	MAY	JUN	JUL	AUG	SEP	OCT	Sum
Abuse	0	2	0	0	3	1	1	1	0	7	3	9	27
Unlawful Activity	0	0	0	0	0	0	0	0	0	0	0	0	0
Malware	3	0	3	5	1	2	1	0	0	2	5	1	23
Reconnaissance	0	0	0	0	0	0	1	0	0	0	2	0	3
Compromised Asset	18	7	8	3	12	8	9	6	6	6	9	19	111
Compromised User	6	5	9	13	12	8	9	9	5	7	1	3	87
Compromised Information	0	0	0	0	0	0	0	0	0	0	0	0	0
Vulnerable Asset	2	5	10	7	5	3	5	8	3	5	5	11	69
Denial of Service	1	0	0	0	2	0	0	0	0	0	1	0	4
Social Engineering	10	9	4	10	19	11	8	11	6	15	17	23	143
Intrusion Attempt	0	0	1	0	0	0	0	0	0	1	0	1	3
Policy Violation	1	0	21	27	10	0	0	0	0	6	1	0	66
Other	4	0	1	1	0	1	0	2	1	2	2	0	14
Sum	45	28	57	66	64	34	34	37	21	51	46	67	

Although the dataset in this paper is limited, we can construct basic predictive models using confidence intervals. These models will improve over time with more data. The top 5 risks per year are in Table 6. For example, the most frequent cause of incidents are low consequence phishing attacks: For the coming year, we expect to see between 85 and 114 attacks with a 90% CI. Copyright violations primarily caused the risk entailing breach to the IT policy, similar to one of the largest categories in the UK incident data [5]. The risk was found unacceptable, and mitigation measures were implemented in March 2017. The effect of the risk treatment can be seen in Table 4 where the number of policy violation drops after April 2017, similar to the Unlawful activity classification in Table 5.

Table 5. Yearly development in incident outcomes

	NOV	DEC	JAN	FEB	MAR	APR	MAY	JUN	JUL	AUG	SEP	OCT	Sum
Abuse	6	3	9	15	10	7	9	10	5	5	2	3	84
Unlawful Activity	0	0	19	27	8	0	0	0	0	1	0	0	55
Malware	13	4	3	3	5	5	2	0	0	4	1	8	48
Reconnaissance	2	2	2	1	5	1	0	0	0	3	2	2	20
Compromised Asset	1	0	2	0	1	0	0	0	0	1	0	0	5
Compromised User	1	0	1	0	2	0	0	0	0	1	0	0	5
Compromised Information	2	1	1	0	0	0	1	1	0	0	2	0	8
Vulnerable Asset	0	0	2	0	0	0	0	0	0	0	0	0	2
Denial of Service	6	5	12	7	8	1	8	4	4	5	5	3	68
Social Engineering	1	2	0	0	2	2	1	1	2	0	1	1	13
Intrusion Attempt	1	1	1	1	2	3	0	0	1	3	2	6	21
Policy Violation	0	0	1	0	0	0	0	0	0	0	0	0	1
Other	1	0	0	1	0	0	0	0	0	1	1	0	4
Negligible/Fixed/Failed	8	10	3	11	21	14	13	20	8	26	27	40	201
Sum	42	28	56	66	64	33	34	36	20	50	43	63	

5   Risk Analysis and Visualization

The previous section presented the numbers and brief analysis of incident occurrences and trends. This section presents a more detailed analysis of the causes and outcomes and how the dataset can reveal data for decision-making. We also present an example of the bow-tie analysis scenario, where we model the case of malware infections. Lastly, we model the incidents using the asset, threat, and vulnerability paradigm for security risks and derive the frequency of occurrence.

Table 6. Confidence Intervals of top five risks occurring per year.

Cause	Outcome	Per year	90%CI Lower	90%CI Upper
Phishing	Negligible/Fixed/Failed Attack	99	85	114
Breach to IT Policy	Copyright/Piracy	52	42	64
Regular User Compromise	Spamming	42	33	53
Open for Abuse	DDoS Outgoing	29	21	39
System Compromise	Trojan	29	21	39

5.1   Cause and Outcome Analysis

This section illustrates the cause and outcome analysis as an initial step into creating a graphical risk representation. We will start by using compromised accounts as an example. Account compromise is one of the most frequent causes of incidents at the NTNU SOC, Table 2. A compromised account is when a company username and corresponding password gets compromised by attackers. Account comprise has previously lead to costly incidents at NTNU, such as the previously mentioned *Silent librarian* campaign caused around 15 incidents (uncertain attribution) and incidents where the network is used as a staging point and is a priority risk to mitigate. During the year of data collection, there were 84 incidents recorded caused by regular user compromise, averaging seven incidents per month. The trend is illustrated in Fig. 4 which shows a peak in February 2017 with 13 incidents caused by compromised accounts and a low in September 2017 with only one. Analyzing the incidents, we find a distribution of outcomes illustrated in Fig. 5. The most frequent outcome of an account compromise is spamming internal users, however, several other outcomes are more severe, with misuse of resources and whaling/CEO fraud attempts bearing the potentially most severe consequences. From an outcome perspective, we can apply this data in the decision process to choose consequence reducing measures to control risk. Looking at the causes of user compromise, Table 3 shows that we only have five incidents where user compromise was the known outcome, where all had been caused by social engineering attacks. Using this approach, we can also reveal areas with uncertainty: for example, the data reveal a lot about the intentions of attackers who use compromised accounts, but the data is lacking on how the accounts are being compromised.

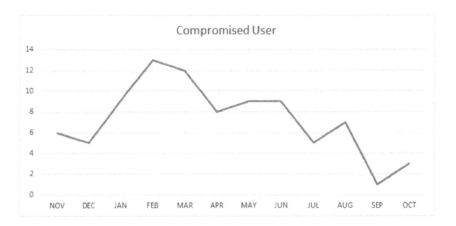

Fig. 4. The figure illustrates the amount of incidents caused by compromised accounts per month.

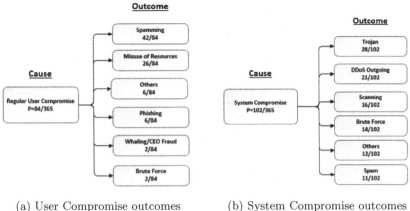

(a) User Compromise outcomes   (b) System Compromise outcomes

Fig. 5. Distribution of outcomes from incidents.

Another case with missing initial causes is the system compromise, which is the "catch-all" category with 102 occurrences. When attackers compromise systems and establish a foothold in the network, the incident outcomes reveal some of their intent, illustrated in Fig. 5. For example, in twenty-eight instances, machines are infected with trojan virus. Typically, the observable action is the call home to the Trojan owner. In thirty instances the infected systems are used as staging points for further attacks to compromise more systems through scanning and brute-force. The data also reveals that twenty-one compromised systems are recruited into botnets and participate in outgoing DDoS attacks on third parties. A thing to note with the system compromise category is that the uncertainty regarding the cause will be reduced with increased detection and forensics capability. The data also reveals that twenty-one compromised systems are recruited into botnets and participate in outgoing DDoS attacks on third parties, which is a problem since outgoing DDoS is illegal and consumes bandwidth from legitimate traffic.

So, if we want to prevent specifically the outgoing DDoS risk, we can analyze it further using the data: In total, we experience sixty-one incidents caused by outgoing DDoS attacks. The distribution of causes are illustrated in Fig. 6, whereas attackers exploiting vulnerable systems caused twenty-nine of the incidents. They are typically vulnerable through missing patches, misconfigured, or natively vulnerable to reflexive and amplification attacks. In this manner, the data reveals a lot about the vulnerability in the organizational networks, and risk controls can be designed to address the vulnerabilities. This analysis is also suited for root cause identification and elimination of security problems. In the following section, we will advance the analysis of the cause and outcome approaches by adding security controls.

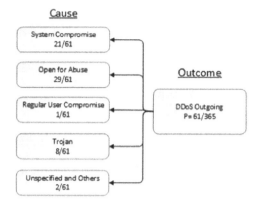

Fig. 6. Distributions of causes for DDoS Outgoing

5.2   Bow-tie Analysis of Malware Infections

This section illustrates the utility of the data for Bow-tie risk analysis (described in Sect. 3.3). Malware infections are at the root of many severe cybersecurity breaches [15] and are present in the current dataset with twenty-five known causes and forty-six known outcomes. This section contains one attack flow model to illustrate the concept.

To populate the bow-tie model, we use the known causes (left side) and outcomes (right side). The unwanted outcome, "Malware infection," is placed in the middle. Further, we have to map out the relevant security controls before we can apply the bow-tie analysis. For the analysis, we are interested in existing preventive controls that reduce the probability of the attack occurring and mitigating controls that reduce the consequence of an incident. Figure 7 illustrates how the classified incident data can be used in bow-tie analysis. All the known causes of malware infections are listed to the left with their known distributions and the outcome distribution to the right. The controls are described at an abstract level not to reveal any defensive capabilities of the SOC. Typically, the bow-tie would make a connection between all causes/outcomes and relevant controls in the figure, but due to the amount and complexity of each incident and the controls involved we created an example instead of including it in the bow-tie.

Figure 8 illustrates how the classified incident data can be used in a bow-tie analysis. Typically, the bow-tie would make a connection between all causes/outcomes and relevant controls in the figure, but we have used only one attack patch as an example for the sake of simplicity. The scenario being analyzed is that an attacker succeeds with a phishing attack, infects the victim with malware, and abuses the machine for DDoS. For simplicity, we focus on only one of the attacks paths for the known causes of malware infections listed to the left together with the preventive controls. The controls are listed in the order the attacker encounters them, e.g., the spam filters will reduce the amount of malicious email that reaches the target (the efficiency is quantifiable). Further-

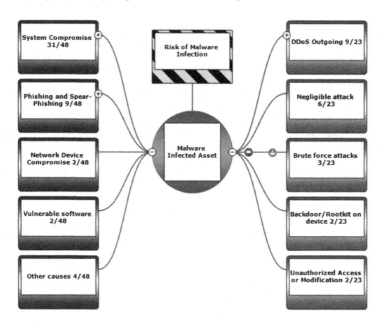

Fig. 7. Bow-tie risk analysis illustrating known causes and outcomes with their respective frequencies for Malware infections in the dataset. Modelled with *BowTieXP CGE Risk Management Solutions*

more, if the email with the malicious link or attachment reaches the target, the security awareness of the user is his/her ability to recognize the phishing attack. If this control also fails and the target opens the malicious link or attachment, the endpoint security is a remaining barrier that can prevent an infection. If all controls fail, the malware infection occurs. The relevant consequence mitigating controls are listed to the right together with the undesirable outcome. The model illustrates all the controls involved both before and after a malware infection, which enables an analysis of both preventive and consequence mitigating controls to identify the weaknesses in the security chain. The frequencies of each event aid the decision-maker in determining whether a risk is unacceptable or not and if new controls should be added to the control chain or existing controls should be strengthened. Furthermore, this type of risk analysis will allow for measurements of control efficiency when implementing new barriers in the security chain.

Fig. 8. Example risk analysis of an Incident derived from the bow-tie.

Table 7. The five most frequent cause and outcome pairs analysed with an asset, threat, and vulnerability model.

Nr	Cause	Outcome	Frq.	Target assets	Threat actors	Motive	Intent	Vulnerability
1	Social Engineering: Phishing	Negligible/ Fixed/Failed Attack	99	Money, Account credentials, Resources	Criminals	Financial, (intelligence)	Unauthorized access, Misuse, Deny access	Human factor
2	Breach to IT Policy by registered user	Copyright or Piracy	52	Resources (bandwidth)	Insiders	Financial, neglect	Misuse	IT policy
3	Regular user compromise	Used to send spam	42	Money, Account credentials, Resources	Criminals	Financial	Unauthorized access, Misuse	Human factor, weak passwords, etc.
4	System resource open for abuse	Exploited in outgoing DDoS attacks	29	Resources (bandwidth)	Opportunists, Criminals	Revenge, Financial	Deny Access	Weak configuration
5	Compromised system	Installed Trojans	28	Company systems, Secrets	Criminals, APT, Chaotic actors	Financial, Political	Unauthorized access, Deny access, Staging point	Vulnerable systems

5.3   Identifying Assets, Threats and Vulnerabilities

The classic ISRA approach from ISO/IEC 27005:2011 [2] advocates to begin the risk assessment process with asset identification and evaluation, before identifying the threats, controls, and vulnerabilities, and inducing the risk. We have generalized the incident data into the more traditional ISRA [17] in Table 7. The model builds on the categorized versions of the incidents and therefore contains some generalizations. The table provides an organizational risk picture for decision-making based on incidents. A more detailed model can be obtained through a more thorough analysis of the incident data. From the dataset, the most frequent risk is phishing attacks, while copyright violations are the second most frequent, and user compromise with internal spamming is the third.

The analysis of the incident data provides a strong indication on what the attackers think that the University's primary assets are (see outcomes, Table 2):

(i) Computing power and resources - for conducting attacks and as a staging point, which is evident from the amount of attacks launched from the network through scanning and brute force attacks. Company accesses are abused to mine resources that are only available through university contracts, such as the Silent librarian campaign, and for the hosting of illegal content.

(ii) Bandwidth capacity - recruited in outgoing DDoS attacks and for illegal file sharing in violation of copyright laws.

(iii) User and admin accounts - Harvested and traded, provides access to company resources, and is used in phishing/spamming. Besides, financial motives are apparent through attempts of CEO frauds/whaling, phishing, and ransomware.

(iv) Information - only seven of the 550 incidents ended in a known information compromise, which indicates that the majority of the attacks that cause incidents aims to exploit other assets at the University. While securing information is essential, the data shows that it is the accesses and resources the University governs that was most interesting for the attackers in the time frame. Using risk nr 1 in Table 7, we see that NTNU is also a frequent target of social engineering campaigns. We deduce that the generic motive behind phishing campaigns is financial as they typically target usernames and passwords, financial data, and other resources. This behavior is typical for cyber-criminals using low-cost social engineering attacks. The table also reveals that the University network is a popular staging point for launching attacks and recruiting resources into DDoS attacks. We attribute this to opportunists and criminal groups. In his Policy note, Chapman [5] attributes many of the DDoS attacks to disgruntled students and staff due to the timings of the attacks. Considering the trends in the incident data, Table 5, provides no support for this being the case at the University.

Regarding vulnerability, the data reveals two specific weaknesses: social engineering attacks and vulnerable systems. So, to reduce incidents, conventional treatments are awareness training and improving the system portfolio and patching routines. Although the dataset is lacking in knowledge about causes for compromised accounts, the primary attack vector against NTNU is social engineering attempts. These attacks typically target account information, and although the incident report states an attack was handled, it is likely unrecorded instances of employees falling for the scam and not reporting. In this case, the dataset allows for hypothesis formulations that can be researched in future projects.

The current threat hype in InfoSec is Nation-state backed groups, or so-called advanced persistent threat (APT). These groups are typically involved in influence operations, sabotage, and cyber espionage [15]. Due to some of the technological research being conducted at universities, they are natural targets of espionage looking to gain a technological advantage. Although the incident data do reveal Trojan activity and more advanced attacks, the dataset does not necessarily reveal advanced persistent threat (APT) activity. The skilled actors are better at hiding their tracks and more data and mature capabilities are needed to conclude that such actors are present in the systems. This is a limitation of the current dataset which can be addressed with forensic capability.

6   Discussion, Limitations, and Future Work

This section discusses the contribution, limitations, and path for future work for each research question proposed in this paper.

6.1   Classifying Incidents

The contribution of this paper is primarily practical: The proposed classification framework contributes to solving the practical problem of quantifying incident data for risk analysis. As we demonstrated in the study, the proposed method enables an overview of incidents that will improve the understanding of the risk landscape at the organization. Although incident reports may vary in format and content, the proposed framework and method has been validated on 550 incidents and should be adaptable for risk quantification at most organizations having InfoSec incident records. The practical implication of the framework is that it enables simple statistical models of risk frequencies and trends, together with graphical modeling in bow-tie diagrams. The approach also facilitates more sophisticated risk analysis to reduce the uncertainty, especially related to frequencies of occurrence and lends itself to the critical incident tool in Root Cause Analysis [9].

This work has also highlighted the ambiguity of the current state of the art regarding incident classification. Although there is an apparent separation between causes and outcomes, it is not considered as a part of current approaches. Considering our results in Table 2, the distinction becomes clear with regards to several classes in the applied framework. We have grouped the level 1 categories on frequencies of occurrence in the cause and outcome in Table 8. This result will depend on three variables: The constituency, how one structures the classification, and how one defines an incident. For example, the organization in our case study is facing a specific risk landscape which places the majority of reconnaissance, intrusion attempts and DoS as primarily outcomes according to our framework. However, the classification could be restructured in future work with a firmer approach to causes and outcomes. All of the incident outcomes related to the three mentioned categories could be restructured, for example, brute force, scanning, and outgoing DDoS are all abuses of resources when categorized as outcomes. Furthermore, the definition of an incident will impact the statistics. Does failed phishing attempts qualify as an incident? In this paper, the answer is yes, because they generated additional work for the SOC and have a measurable consequence in terms of time spent managing user inquiries. However, this may not be the case for other SOCs and industries. Incidents managed with automation also represent a challenge for incident registers. For example, one malware incident may be handled manually and logged, while another incident is dealt with automatically with endpoint protection and does not make it into the register. The incident response community should deal with these problems since the incident statistics are essential in reporting both internally and to the public.

All models are simplifications of reality, and as mentioned in the introduction, an incident can consist of a chain of causes with multiple adverse outcomes, rather than just one cause and outcome as implied in this framework. We recognize this issue, and a more sophisticated approach should be considered in cases where more detailed information is needed. An approach model containing multiple causes, such as a root cause analysis-approach [9] or the Lockheed Martin

Table 8. Overview of applied categories sorted on Cause and Outcome

Primarily Cause	Both	Primarily Outcome
Compromised Asset	Abuse	Unlawful activity
Compromised User	Malware	Reconnaissance
Vulnerable Asset	Other	Compromised Information
Social Engineering		Denial of Service
Policy Violation		Intrusion Attempts

Cyber Kill Chain can be adapted to improve the model. A more cost-efficient strategy would be to reserve the thorough analysis to more severe incidents and to do simple quantification of day to day attacks. Another limitation is with the method in this paper: while it is likely that the level 1 classifications can be generalized to most organizations and industries, the level 2 classifications should be tailored to the organization planning to use them. However, both the level 1 and 2 classifications as proposed should provide a starting point for incident classification. A suggestion for future work would be to work on a common framework for a higher detail incident analysis based on traditional ISRA.

Furthermore, all of the categorization done for this paper was done by an analyst, which adds subjectivity in the analysis. We attempted mitigating this issue by applying firm categorizations, outlining rules for categorizations, and adapting the framework as needed. It is a challenge to keep the categories unambiguous and prevent overlap between them. For example, outgoing DDoS is both a DDoS attack and abuse of network infrastructure. Our main categories were developed from the best practice and is similar to those applied in the Jisc SOC [5], this same ambiguity is seen across frameworks. A path for future work is to propose a framework for incident classification using generic risk classifications as a starting point. It is clear from the Table 2 some of the categories are more likely to be a cause of an incident than the outcome and vice verse. Refining causes and outcomes with risk quantification as the goal could assist in improving risk management of cybersecurity risks.

The incident analysis and classification is a time consuming and repetitive job, which makes automation another path for future work. Depending on the incident record system, the process of automatically classifying incidents could be developed by retrieved the data, produce the dataset for machine learning, develop identifiers for each category, and develop the algorithm for incident classification and risk quantification.

6.2   The Risk Picture

All models are simplifications of reality, and as mentioned in the introduction, an incident can consist of a chain of causes with multiple adverse outcomes, rather than just one cause and outcome as implied in this framework. We recognize this issue, and a more sophisticated approach should be considered in cases where more detailed information is needed. An approach model containing multiple causes, such as a root cause analysis-approach [9] or the Lockheed Martin *Cyber Kill Chain* can be adapted to improve the model. A more cost-efficient strategy would be to reserve the thorough analysis to more severe incidents and to do simple quantification of day to day attacks. Another limitation is with the method in this paper: while it is likely that the level 1 classifications can be generalized to most organizations and industries, the level 2 classifications should be tailored to the organization planning to use them. However, both level one and two classifications as proposed should provide a starting point for incident classification. A suggestion for future work would be to work on a common framework for a higher detail incident analysis based on traditional ISRA.

Furthermore, the categorization done for this paper was done by an analyst, which adds subjectivity in the analysis. We attempted mitigating this issue by applying firm categorizations, outlining rules for categorizations, and adapting the framework as needed. It is a challenge to keep the categories unambiguous and prevent overlap between them. For example, outgoing DDoS is both a DDoS attack and abuse of network infrastructure. Our main categories were developed from the best practice and are similar to those applied in the Jisc SOC [5], this same ambiguity is seen across frameworks. A path for future work is to propose a framework for incident classification using generic risk classifications as a starting point. It is clear from the Table 2 some of the categories are more likely to be a cause of an incident than the outcome and vice verse. Refining causes and outcomes with risk quantification as the goal could assist in improving risk management of cybersecurity risks.

The incident analysis and classification is a time consuming and repetitive job, which makes automation another path for future work. Depending on the incident record system, the process of automatically classifying incidents could be developed by retrieved the data, produce the dataset for machine learning, develop identifiers for each category, and develop the algorithm for incident classification and risk quantification.

6.3 Risk Visualization

We applied the bow-tie diagram to illustrate the utility of the dataset. Applying bow-tie diagrams to the data allowed for the construction of simple attack flows with frequencies of occurrence both for cause and outcome for each incident category. The strength of this approach is that it is easy to understand and communicate. The risk visualization also allows security control modeling and measurement of control efficiency. The drawback of the diagrams is that they can be an over-simplification of reality in cases of severe risks. In these cases, more sophisticated modeling techniques can be used, such as *event trees* and *attack trees*. More extended attack flow diagrams with multiple causes and outcomes is also a possibility. Another path for improvement is to work on the bow-tie diagrams for the more severe risks and research ways of measuring control efficiency and integrating them into the risk assessment model. The risk model should also include working with loss estimates for the identified outcomes. Adapting

Kuypers et al. [12] approach for differentiating incident consequences based on time spent handling the incident is a start. However, only considering the cost of time spent handling the incident represents a too narrow view on consequences, as there are also possibilities for production loss, asset damage, legal fines, and reputation/competitive advantage to consider. The incident impacts to each of these areas can be estimated by applying the approach proposed by Seiersen and Hubbard [10]. Another limitation with the information presented in this paper regarding the asset, threat, and vulnerability are generalizations from the incident data. More specific information on each is present in the incident data but not reported in this paper. A limitation regarding threat actors is that we can not know who they are because of the *attribution problem* [15]. The threat actors are categorized on perceived motivation using the approach proposed by Potter [13] and Wangen et al. [17, 18].

7   Conclusion

With this paper, we have highlighted some key issues faced when analyzing incident data and proposed a classification approach to smooth the transition from the incident report to risk quantification and analysis. Our proposed method and framework was anchored in a two-level approach based on established incident classifications and expanded when necessary. The framework was scoped to classify incident causes and outcomes for quantification. By applying the method to a case, we were able to create an empirical risk picture for the University, including all the known causes and outcomes of incidents. This study found that one will not get a complete risk picture from analyzing the incident data alone, but it will provide valuable insight into critical issues the organization faces: According to the data, the risk picture for NTNU contained a range of cyber attacks, such as social engineering attempts, vulnerable/compromised devices, malware infections, and DDoS. The most frequent threat was identified as social engineering attempts, including phishing, spear phishing, and whaling/CEO frauds. Compromised assets and users made out the second and third most frequent causes of incidents. About 2/5 of the InfoSec incidents were resolved without any observable adverse outcome. Abusing the University infrastructure through outgoing DDoS attacks, spamming, and copyright violations are the three most frequent outcomes in the data set. Although 12 months of data is a short period, we have demonstrated how to apply the proposed approach to study trends in both the cause and the outcome from incidents.

Furthermore, the proposed incident analysis method has merit when integrated with the bow-tie analysis, as the model fully utilizes both the cause and outcome statistics. Quantifying incidents allows for predictions, but it also enables the risk analyst to measure the treatment effect over time. In cases such as compromised accounts, we also applied the analysis to detect knowledge gaps, whereas the data revealed that there was little knowledge as to how accounts were compromised.

Acknowledgements. The NTNU digital security section and SOC consisting of Christoffer Vargtass Hallstensen, Frank Wikstrøm, Harald Hauknes, Hans Åge Marthinsen, Vebjørn Slyngstadli, Gunnar Dørum, Lars Einarsen, and Stian Husemoen. Vivek Agrawal and the anonymous reviewers for help with quality assurance.

References

1. Common taxonomy for law enforcement and the national network of csirts, version 1.3. Technical report, ENISA and Europol E3 (2017). https://www.europol.europa. eu/publications-documents/common-taxonomy-for-law-enforcement-and-csirts
2. Information technology, security techniques, information security risk management (ISO/IEC 27005:2011)
3. Reference incident classification taxonomy: Task force status and way forward. Technical report, ENISA, January 2018
4. Bernsmed, K., Frøystad, C., Meland, P.H., Nesheim, D.A., Rødseth, Ø.J.: Visualizing cyber security risks with bow-tie diagrams. In: Liu, P., Mauw, S., Stølen, K. (eds.) GraMSec 2017. LNCS, vol. 10744, pp. 38–56. Springer, Cham (2018). https://doi.org/10.1007/978-3-319-74860-3_3
5. Chapman, J.: How safe is your data? cyber-security in higher education. HEPI Policy Note, 12 April 2019
6. Edwards, B., Hofmeyr, S., Forrest, S.: Hype and heavy tails: a closer look at data breaches. J. Cybersecur. **2**(1), 3–14 (2016)
7. Florêncio, D., Herley, C.: Sex, lies and cyber-crime surveys. In: Schneier, B. (ed.) Economics of Information Security and Privacy III, pp. 35–53. Springer, New York (2013). https://doi.org/10.1007/978-1-4614-1981-5_3
8. Hansman, S., Hunt, R.: A taxonomy of network and computer attacks. Comput. Secur. **24**(1), 31–43 (2005)
9. Hellesen, N., Torres, H., Wangen, G.: Empirical case studies of the root-cause analysis method in information security. Int. J. Adv. Secur. **11**(1&2), 60–79 (2018)
10. Hubbard, D.W., Seiersen, R.: How to Measure Anything In Cybersecurity Risk. Wiley, Hoboken (2016)
11. Kjaerland, M.: A taxonomy and comparison of computer security incidents from the commercial and government sectors. Comput. Secur. **25**(7), 522–538 (2006)
12. Kuypers, M.A., Maillart, T., Pate-Cornell, E.: An empirical analysis of cyber security incidents at a large organization. Department of Management Science and Engineering, Stanford University, School of Information, UC Berkeley 30 (2016)
13. Potter, B.: Practical threat modeling. Login **41**(3) (2016). https://www.usenix. org/publications/login/fall2016/potter
14. Romanosky, S.: Examining the costs and causes of cyber incidents. J. Cybersecur. **2**(2), 121–135 (2016)
15. Wangen, G.: The role of malware in reported cyber espionage: a review of the impact and mechanism. Information **6**(2), 183–211 (2015)
16. Wangen, G., Brodin, E.Ø., Skari, B.H., Berglind, C.: Unrecorded security incidents at NTNU 2018 (Mørketallsundersøkelsen ved NTNU 2018). NTNU Open Gjøvik (2019)
17. Wangen, G., Hallstensen, C., Snekkenes, E.: A framework for estimating information security risk assessment method completeness. Int. J. Inf. Secur. **17**, 1–19 (2017)
18. Wangen, G., Shalaginov, A., Hallstensen, C.: Cyber security risk assessment of a DDoS attack. In: Bishop, M., Nascimento, A.C.A. (eds.) ISC 2016. LNCS, vol. 9866, pp. 183–202. Springer, Cham (2016). https://doi.org/10.1007/978-3-319-45871-7_12

Poster Support for an Obeya-Like Risk Management Approach

Stéphane Paul[1]([⊠]) [iD] and Paul Varela[2]([⊠]) [iD]

[1] Thales Research and Technology, 91767 Palaiseau, France
stephane.paul@thalesgroup.com
[2] Thales Secure Communications and Information Systems, 92230 Gennevilliers, France
paul.varela@thalesgroup.com

Abstract. Lean management is trendy. This trend is also reaching risk management. It has become very concrete in France following the EBIOS-Risk Manager method publication by the French National Agency for cybersecurity (ANSSI) in October 2018. However, if the new method fosters an agile approach of risk management, it does not provide the tools to support the mandated brainstorming workshops. In this paper we propose a set of A0 posters (and A5 cheat-sheets) to support the efficient and user-friendly organisation of the EBIOS-Risk Manager brain-storming sessions. The workshop participants are given sticky notes and felt pens to actively contribute to the data collection work. A facilitator helps organise the emergence of contributions. This approach is inspired from the Japanese Obeya form of project management, with the goal of making risk management simple, dynamic and attractive, or in one word, fun!

Keywords: Risk management · Agile · Collaborative · Workshops · Brainstorming · Posters · Sticky notes (Post-Its®) · EBIOS

1 Introduction

In Oct. 2018, the French National Agency for Cybersecurity (ANSSI) published a new version of the EBIOS risk management method called EBIOS-Risk Manager [1]. The first version of the EBIOS method dates back to 1995. It was significantly updated in 2004 and 2010. EBIOS-2010 established itself as the main risk management method used in France. The new version of the method brings some significant changes, amongst which the following:

- It explicitly targets populations beyond the classical cybersecurity experts, including company directorates, risk managers, Chief Information Security Officers (CISOs), Chief Information Officers (CIOs), and business/operational experts, such as Architects (ARCs), Product Line Architects (PLAs), Design Authorities (DAs) and System Engineering Managers (SEMs).
- It mandates securing by conformity, prior to securing by scenario. Securing by conformity means that a Minimal Set of Security Controls (MSSC), based on best practices,

© Springer Nature Switzerland AG 2019
M. Albanese et al. (Eds.): GraMSec 2019, LNCS 11720, pp. 155–185, 2019.
https://doi.org/10.1007/978-3-030-36537-0_8

is selected prior to risk identification. Then risks are identified taking into account existing controls. ANSSI's hypothesis is that accidental events should normally be covered by the MSSCs, so that the analysis can focus on malevolence. In other words, only a few incident scenarios[1] should be necessary, either to prove that the solution is secure, or to highlight some rare holes in the system.

- It is run as a set of ½-day workshops (i.e. brainstorming sessions), with 2 to 4 participants per session.
- It is an agile approach, providing quick results for decision-makers. Typically it is possible to start outputting grosgrain risks after only three workshops, corresponding to 1½ days work. To go in depth, it is possible to iterate on the workshops. It is also recommended to iterate through operational and/or strategical cycles, to keep the system in secure conditions throughout its lifecycle. The operational cycle deals with fast changing facts, e.g. vulnerabilities. The strategic cycle deals with slower changing facts, e.g. system missions.
- It is configurable. It is not required to run all five workshops in sequence. The choice to run a workshop depends on the team objectives.
- Its scope is extended to include the ecosystem, a.k.a. system stakeholders. The assumption here is that many attacks do not target directly the system, but first target a stakeholder (e.g. a sub-contractor), then move laterally.

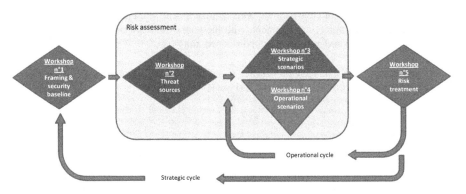

Fig. 1. The five workshops of the EBIOS-Risk Manager method

The novelty of EBIOS-Risk Manager that is of interest to us in this paper relates to the organisation of the risk management work in the form of workshops. The goal of this paper is not to present or promote the EBIOS-Risk Manager method. However, since it is only available in French at the day of writing this paper, we provide some insight and personal translation to allow the reader to understand the relevance of our work. To be brief, let us just say that, as pictured in Fig. 1, the EBIOS-Risk Manager method proposes five workshops called: (1) Framing and security baseline; (2) Risk sources; (3) Strategic scenarios; (4) Operational scenarios; and (5) Risk treatment.

---

[1] According to ISO, an incident scenario is the description of a threat exploiting a certain vulnerability or set of vulnerabilities in an information security incident.

The method specifies the objectives of each workshop, the expected attendees, the expected outputs, and how to proceed. However the guidance is technical in the sense that it typically specifies what data to collect, and how to assess/classify it. The method does not guide on how to organise and conduct the workshop, e.g. how to interact with the participants, collect and document the information, or reach a consensus between the participants.

In this paper, we propose a set of A0 posters (and A5 cheat-sheets) to support the efficient and user-friendly organisation of the EBIOS-Risk Manager brain-storming sessions. A facilitator helps organise the emergence of contributions. This approach is inspired from the Japanese Obeya form of project management, with the goal of making risk management simple, dynamic and attractive, or in one word, fun!

Section 2 describes the content of the different poster templates supporting the five EBIOS-Risk Manager workshops. Section 3 discusses scalability, method efficiency and briefly explains how a final cybersecurity report can be generated following the workshops. The conclusion recalls the history of the creation of the posters, and how their maturity was increased through a series of case studies; we also provide some hints on the way forward. Extensive appendixes provide examples of how the posters have already been used on real live case-studies.

2 A Set of Posters to Support Cybersecurity Risk Identification, Assessment and Treatment

In this section we present our material to organise the EBIOS-Risk Manager [1] brain-storming sessions in a user-friendly way. We will proceed workshop per workshop. However, we start by explaining some organisational elements that are common to all brainstorming sessions.

EBIOS brainstorming sessions typically involve between 2 and 4 persons, in addition or including the facilitator. Each participant is given a single A5 cheat-sheet that recalls the workshop objectives and provides some hints as to how the session is going to be run. Sheets may define terms, scales, small knowledge bases, etc. In practice, we noted that participants spent very little time reading the cheat-sheets. However, the cheat-sheets have a reassuring psychological effect: "*I can always refer to the sheet if I get lost*". The cheat-sheets we developed are not further discussed within this paper.

Filling posters with sticky notes during brainstorming sessions is the most productive part of the work. However, our approach also requires significant back-office work to produce "clean" versions of the posters. By clean, we mean that the hand-written sticky notes collected during the workshop are typed in using PowerPoint. During back-office work, care is taken to place the electronic copies of the sticky notes at the same place as they were set during the workshop, to leverage the visual memory of the participants. In between workshops, the clean poster(s) are sent to the participants by email, for validation. At this stage, feedback is generally very low. This is not an issue because before beginning a workshop, the poster(s) of the previous workshop are submitted for live review. We noted that the reviews classically require no more than 5 to 10 min. It is quite frequent that some small amount of corrections and addendums are performed during the reviews. The facilitator responsible for the back-office work may also have spotted some inconsistencies or incompleteness, which he should bring up for discussion during the reviews.

2.1 Workshop n°1: Framing and Security Baseline

Framing and Security Baseline is the first workshop of the EBIOS-Risk Manager method. Its goals are to frame the system-under-study, identify its missions, security needs, and start building a cybersecurity engineering strategy. Its participants should be a top manager, a domain expert, the CISO and the CIO.

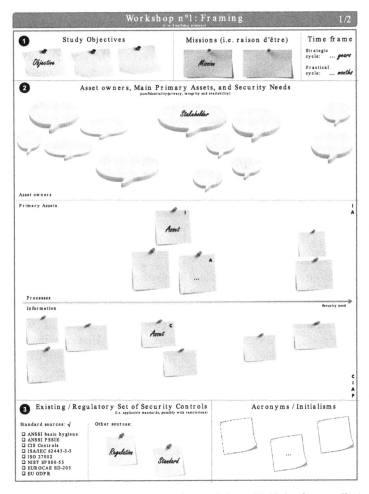

Fig. 2. First poster template supporting workshop n°1 (Color figure online)

The expected outputs are some framing elements, e.g. study objectives, roles and responsibilities, the domain and technical perimeter, including business/operational assets, the feared (a.k.a. undesired) events and their severity, and the Minimal Set of Security Controls (MSSC) to be applied. To support this workshop, we propose two mandatory A0 posters, plus some optional A0 posters. The first poster that we propose is pictured in Fig. 2.

As a first step, the poster allows to capture the study objectives, the missions of the system-under-study and the time frame for the strategic and operational/practical cycles. These cycles refer to the time governing the workshop iterations (cf. Fig. 1), to ensure maintenance in secure conditions during the whole system lifecycle.

As a second step, the poster allows to capture the asset owners, primary assets (a.k.a. business or operational assets), and their security needs. The asset owners are represented by comic strip speech bubbles to underline the fact that these stakeholders have their word to say. Blue sticky notes are used for primary assets. They are split in two: processes and information, as defined in ISO 27005 Annex B [2], and arranged on either side of a security need axis. The security need axis is represented by a horizontal arrow whose colour spans from green, meaning low security need, to red, meaning high security need. It can be used as an asset valuation [2]. Above the axis, the area is earmarked for valued processes, below for information. The security needs are usually expressed in terms of Confidentiality (C), Integrity (I) or Availability (A), to which we have added Privacy (P) due to the recent entering in force of the European General Data Protection Regulation (GDPR). We recommend marking the security need by a capital letter on the upper right part of each sticky note.

In practice, we noted that participants care more about the relative position of a sticky note, rather than its absolute position, because it allows setting priorities, so that the risk management process can quickly focus on the most important primary assets. If a primary asset has multiple security needs, but with a different sensitivity for each need, then we recommend the use of multiple sticky notes. E.g., if the integrity and availability need of an information is high, whilst its confidentiality need is low, then two post-is should be created, one with the "IA" marking (for integrity & availability) on the red part of the security need axis, and one with the "C" marking (for confidentiality) on the green part of the security need axis.

As a third and last step, the poster allows capturing the name of security control standard(s) which may be mandated on the study. To ease the capture, some common international standards are already listed, but more can be added. At this stage only the names of the standards are listed. In the next poster, some space is dedicated to listing already existing or specified security controls. Finally, since all businesses come with their specific jargon, space is also given to define some key acronyms or initialisms. A filled example of this poster is given in Fig. 12 (appendixes).

The second A0 poster that we propose to support the first EBIOS-Risk Manager workshop is pictured in Fig. 3. Since this poster follows the previous one (cf. Fig. 2), it directly starts with step n°4, as indicated in the black circle at the upper left side of the poster. Step n°4 is dedicated to the identification of supporting assets. The poster offers three areas dedicated respectively to organisational assets (i.e. personal, organisation's structure), Information Technology assets (i.e. hardware, software, network), and physical assets (i.e. premises and infrastructure) as defined in ISO 27005 Annex B [2].

The fifth and last section of the poster relates to existing or already specified security controls. A filled example of this poster is given in Fig. 13 (see appendixes).

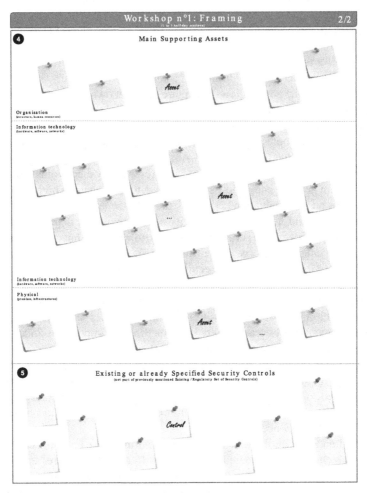

Fig. 3. Second poster template supporting workshop n°1

Some projects may need to assess the implementation status of the different security controls, in particular projects that build on an existing systems or infrastructures. Typically, one may need to assess if the existing and/or specified security controls are currently fully implemented, partially implemented or not implemented. To support this assessment work, we have developed an additional poster presented in Fig. 11 (see appendixes). Note however that this poster has not yet been used in any real life case-study, so its maturity may be significantly lower than the poster templates presented in this section.

In the above, we have not yet explained how the existing or already specified security controls are defined. In practice, the baseline may be defined by the customer, regulation, standards, engineering best practices, or it may be derived through a process known as *System Security Categorisation* in NIST SP 800-64 [3] and in the Thales Engineering Baseline [4]. To support System Security Categorisation, we propose an optional A0 poster that allows for the assessment of the severity of the impacts of the feared events. This optional poster template is shown in the appendixes, in Fig. 10, with a filled in example in Fig. 14.

Overall, four A0 poster templates are proposed to support the method's first workshop, of which two poster templates are optional.

2.2 Workshop n°2: Risk Sources

The second workshop of the EBIOS-Risk Manager method is called Risk Sources. Its goals are to identify who or what may jeopardise the primary assets identified during the previous workshop, and to what ends. Its participants should be a top manager, a domain expert, the CISO and, if possible, a Threat Intelligence (TI) expert. The expected output is a prioritised map of the risk sources, and their objectives. To support this workshop, we propose a single A0 poster, presented Fig. 4.

As a first step, the poster allows to capture risk sources (green sticky notes), classified as: (i) intentional external; (ii) intentional internal; (iii) accidental, whether human or natural, internal or external to the system-under-study. The risk sources can be sorted by relevance, where relevance is generally assessed taking into account resources available to the risk source, the risk source motivation and its activity history (i.e. precedents). A column on the far left, allows for the capture of rejected risk sources.

As a second step, the most relevant risk sources are selected and repeated using green sticky notes on the lower part of the poster. Below each risk source are represented associated adverse objectives. Once again, the objectives can be sorted by relevance. A dashed horizontal line cuts through the area dedicated to the adverse objectives. All adverse objectives above that line will be studied in the following workshops; the study of the adverse objectives below the dashed horizontal line will be delayed until the next risk management cycle. The leftmost column allows for the capture of rejected adverse objectives. A filled example of this poster is provided in Fig. 16. If more than four risk sources need to be considered, we propose a poster extension (see Fig. 15). In terms of scalability, the method recommends dealing with an average of 3–6 adverse objectives per cycle, so as to keep it manageable during brainstorming.

2.3 Workshop n°3: Strategic Scenarios

The goal of the third workshop is to describe high-level scenarios stating how the previously identified risk sources (cf. Sect. 2.2) can attack the system-under-study. ANSSI asserts that a significant part of cybersecurity attacks do not target the system directly, but first target some system stakeholder, and then move laterally to attack the system. Thus, before describing strategic scenarios, EBIOS-Risk Manager mandates the mapping of the ecosystem, i.e. external stakeholders interacting with the system-under-study, and the identification of critical stakeholders, i.e. those most likely to be targeted by a risk source. The workshop participants should be a domain expert, an architect, the CISO and, if possible, a cybersecurity expert. The expected outputs are a mapping of

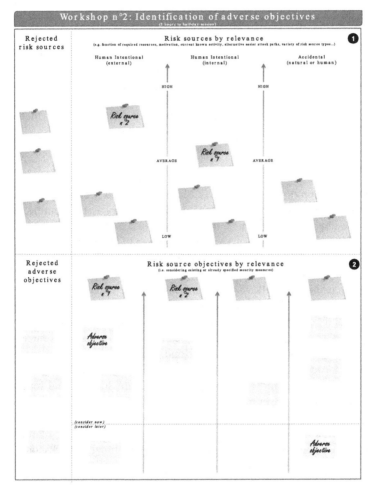

Fig. 4. Poster template supporting workshop n°2 (Color figure online)

the ecosystem, a list of critical stakeholders, a prioritized list of strategic scenarios and a proposal of complementary security controls. Feared events may also be studied during this workshop if they were not studied during workshop n°1.

To support this workshop, we propose two mandatory A0 posters, plus one optional A0 poster. The workshop n°3 optional poster is identical to the workshop n°1 optional poster (cf. Figs. 10 and 14), to allow for the assessment of the severity of the impacts of the feared events. It is therefore not further discussed herein.

In terms of wording, we have introduced the term *risk* instead of the EBIOS *strategic scenario* expression. It is our feeling that this is a more natural concept for non-cybersecurity experts, and up to now, the case-studies that we have run have not shown any distortion due to this wording simplification.

The first poster template that we propose is pictured in Fig. 5. The top part of the poster is dedicated to the mapping of the ecosystem. The schema presents two scales: (a)

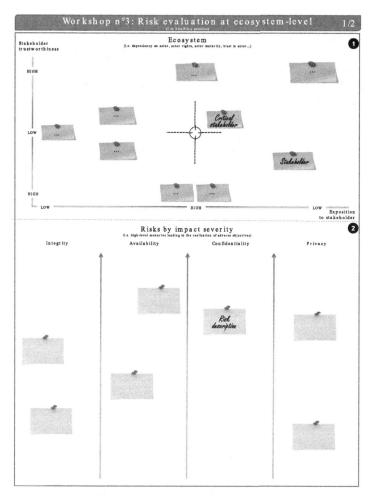

Fig. 5. First poster template supporting workshop n°3 (Color figure online)

horizontally, the scale relates to the system's exposition to the stakeholder; (b) vertically, the scale relates to the trustworthiness of the stakeholder. Initially, the schema is a bit difficult to get used to because the axes are bidirectional. This complex layout, which was recommended to us by ANSSI, has several ergonomic advantages: the stakeholders who are untrustworthy and to which the system is significantly exposed are pictured in the *centre of the diagram*. Naturally they are at the *centre of attention*. Hopefully such stakeholders should be scarce. A filled example of this poster is provided in Fig. 18.

In terms of colour, we reused the green colour to capture the ecosystem. This colour is similar to the colour used for the risk sources during workshop n°2. We do not mean by that the ecosystem should be considered as risk sources but we do mean that those stakeholders are potential threat actors used by a risk source to perform the attack. The difference between threat source (a.k.a. risk source) and threat actor is well captured in the HMG IA standard [5, 6].

The lower part of the poster (cf. Fig. 5) is used to capture high-level descriptions of the risks. Pink sticky notes are used here once again, as it was done for the risk source objectives in workshop n°2 (cf. Sect. 2.2). Indeed, the risks described herein are possible attack paths, which describe how the risk sources will proceed to reach their objective(s), at a high level. The facilitator should make sure that at least one scenario is described for each adverse objective identified during the previous workshop.

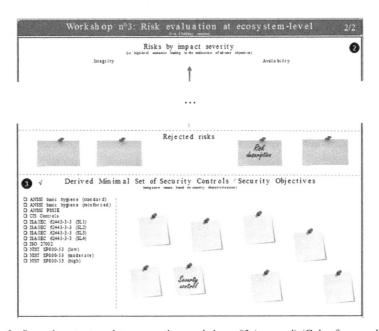

Fig. 6. Second poster template supporting workshop n°3 (cropped) (Color figure online)

In this diagram, the high-level risk descriptions are sorted by impact severity, according to a vertical severity axis, from low impact severity at the bottom and in green, to high severity at the top and in red. As for the previous diagrams, it is again a relative positioning that is important, in order for the risk assessment team to rapidly focus on the most severe risks. A filled example of this poster is given in Fig. 18.

The second poster template supporting workshop n°3 (Fig. 6) extends the previous poster (the upper part, similar to the previous poster, was cropped out), and also allows for the capture of rejected risks. Capturing rejections asserts that those risks were considered, and not simply forgotten.

Normally risk descriptions will be rather extensive: for this section, we recommend large sticky notes. In the follow-up, risks can be numbered, and referred to by their number. A special sheet has been prepared to serve as risk register (cf. Fig. 17).

Finally, the lower part of the poster supports the last step of the EBIOS-Risk Manager workshop n°3, i.e. the identification of additional security controls to deal with intrinsic vulnerabilities of critical stakeholders. As in workshop n°1, yellow sticky notes are used to capture the security measures. And here again, the poster template presented in Fig. 11

(see appendixes) can be used to assess the implementation status of the ecosystem-related security controls.

2.4 Workshop n°4: Operational Scenarios

The goals of the fourth workshop are to describe *how* the strategic scenarios will be realised, using an ad-hoc version of the Lockheed Martin Cyber Kill Chain® [7], and assess their likelihood. Thus, this workshop requires a good knowledge of the supporting assets (as captured during workshop n°1) and their vulnerabilities. The workshop participants should be the CISO, the CIO and preferably, a cybersecurity expert. The expected output is a list of operational scenarios with their likelihood of occurrence. To support this workshop, we propose an A0 poster template to capture the cyber kill-chains (cf. Fig. 7), plus another poster to synthetize the risks. This last poster is in fact shared between workshops n°4 and n°5, the top part being filled in during workshop n°4, and the bottom part during workshop n°5.

The poster template to capture the cyber kill-chains is one of our most complex posters, but when the brainstorming participants reach this workshop, they have become familiar with the poster principles; none of the participants we had on our case-studies seemed destabilised by the poster complexity.

The main structuring element of the poster is an ad-hoc version of the Lockheed Martin Cyber Kill Chain® comprising 5 steps, instead of the 7 steps of the original version. As step representation we used a symbol resembling the *task definition* icon standardised in SPEM [8]. Our 5 steps are: (i) External reconnaissance; (ii) Intrusion; (iii) Internal reconnaissance; (iv) Move laterally; and (v) Exploitation. The general idea is that the brainstorm participants detail the way an attacker can achieve his objective going through the five steps listed above. This description should at least encompass (from top-down on the poster):

- The system mode, because an attack scenario depends on the system's state and mode. E.g., an attack which is possible during system development will probably not be possible during system operation. A specific space is earmarked above the Cyber Kill Chain to capture this system mode information.
- The estimated likelihood of the scenario. This poster shows the likelihood captured as a cumulative percentage just below the Cyber Kill Chain, but it is of course possible to use a qualitative scale instead, leveraging a colour-coding mechanism.
- The risk source or risk actor involved in each part of the attack scenario. They are captured using green coloured sticky notes.
- The existing security controls that may prevent the attack from happening, protect against it, or detect and respond to the attack. Existing security controls are captured using yellow coloured sticky notes located above the attack scenario description.
- The description of the attack scenario itself. E.g., reverse-engineering hardware equipment, then installing a root kit. This is done using red coloured sticky notes.

The detailed descriptions provided on this poster should correspond to at least one high level risk description captured during workshop n°3. To establish the link, the reference to the high level risk description is recalled on the upper left part of the poster, using a pink sticky note.

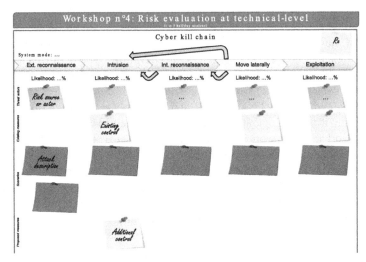

Fig. 7. Poster template to support capturing cyber kill-chains, workshop n°4 (cropped) (Color figure online)

According to the EBIOS-Risk Manager method, risk treatment is performed during workshop n°5. However, when we run this workshop for our case-studies, we witnessed a natural behaviour of the participants to immediately provide countermeasures. Thinking about an attack scenario is a strenuous mental activity. Splitting the work in a description activity during workshop n°4, followed by a treatment activity during workshop n°5 is inefficient and mentally exhausting. We therefore extended our poster to allow for the capture of additional security measures, which will eventually be reviewed during the risk treatment workshop (n°5). The additional security measures are captured using yellow coloured sticky notes located below the attack scenario description. At this stage, these additional security measures are obviously not (yet) taken into consideration when computing the scenario likelihood. The Cyber Kill Chain poster template should be used as often as required to describe all relevant attack scenarios. The work mandated by the EBIOS-Risk Manager method stops here for workshop n°4. A filled example of this poster is given in Fig. 19 (appendixes).

From our perspective, we feel that it is important to conclude this workshop by a synthesis of the inherent (a.k.a. initial) risks – whereas it is part of workshop n°5 according to the EBIOS method. The reason for our position is that at this stage we have both the severity and the likelihood of the scenarios, allowing for the positioning of the risks on a risk aversion matrix. To support this synthesis of inherent risks, we propose a new poster, illustrated in Fig. 8.

This poster is split in two identical sector representations. Each schema presents two scales: (a) horizontally, the risk likelihood scale; (b) vertically, the risk severity scale. During this workshop, we only fill in the top part of the poster, related to inherent risks.

During the next workshop, dedicated to risk treatment, we will synthetize the residual risks. The poster will then offer a comprehensive view of risks, before and after treatment.

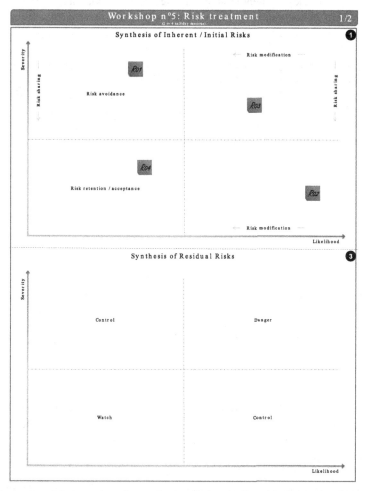

Fig. 8. Poster template to capture the synthesis of inherent & residual risks (workshops 4 & 5) (Color figure online)

To help beginners, the poster also provides risk management hints, in support of workshop n°5. For example, on the top part, the most obvious risk treatment options are listed within the different sectors. Typically, if a risk is very likely and severe, one should essentially consider risk modification, to reduce risk likelihood, or risk sharing to reduce risk severity. By contrast, if the risk is in the bottom-left part of the diagram, one should probably consider risk acceptance/retention.

2.5 Workshop n°5: Risk Treatment

The objectives of workshop n°5, called Risk Treatment, are to synthetize the inherent risks, define a risk treatment strategy, derive the corresponding security measures, integrating them in a continuous improvement plan, and assess/manage the residual risks.

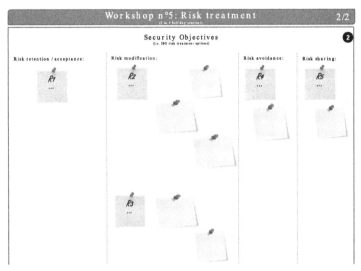

Fig. 9. Poster template to support risk treatment as part of workshop n°5 (cropped) (Color figure online)

The workshop participants should be, as for workshop n°1, a top manager, a domain expert, the CISO and the CIO. To support this activity, we propose a final A0 poster dedicated to risk treatment (cf. Fig. 9). The synthesis of residual risks is performed on the poster already presented as part of workshop n°4, cf. Fig. 8.

Our poster proposal for risk treatment is pretty straightforward. It simply presents four columns, corresponding to the four risk treatment options proposed in ISO 27005 [2], i.e. risk retention/acceptance, risk modification, risk avoidance, and risk sharing. In each column, we propose to reference a risk scenario, using a pink or red sticky note, and then, except for the risk retention option, list the proposed additional security measures, using yellow sticky notes. A filled example of this poster is given in Fig. 20.

Risk treatment involves cost-benefit analysis for different risk treatments. To make rational decisions, one may for example need to know what insurance policies are available on the market for the particular identified risks. As workshops are rather short, a thorough economic analysis of treatment alternatives is not feasible during the allotted time; it must be prepared in advance. This statement is also valid for all previous workshops: the brainstorming and poster-based approach is not a magical wand. To be efficient, the workshops need to be prepared and relevant data collected beforehand.

3   Discussion

3.1   Scalability

Our approach is meant to be use during the project bid phase and/or during the early development phases of the project. During the early development phases of the project, the brainstorming sessions can be organised with or without customer representatives, possibly both, thus exercising two iterations.

The approach has obvious limitations in terms of scalability. To start with, the posters offer a limited surface to stick sticky notes. Next, people in brainstorming sessions can only take so much information at a time. This is clearly intrinsic to the method. For example, for workshop n°2, ANSSI suggests to limit the number for selected threat source objectives to something between 3 and 6. Finally, as the complexity grows, the back office work grows to disproportionate amounts, typically to print and manipulate the A0 posters.

Thus, our recommendation is to realise one or two iterations with this approach, to gain a collaborative momentum and establish a high-level consensus on the risk management priorities and, then switch to more traditional software tools. ANSSI has launched an accreditation process for tool vendors: eight French companies and one Dutch company have already registered [9]. It will therefore not be a problem to find a computer tool to edit the collected data in a more formal framework.

3.2   Methodological Evaluation

It is difficult for us to objectively discuss the efficiency of the method, as it is rare that people run two parallel risk assessments on real industrial programmes, just to compare the results.

However, in our case, our interlocutors on the VLLAM/UTM case-study (see Sect. 6.1) had already run a cybersecurity risk assessment with the Thales Digital Factory. The feedback we collected is that our approach was interesting in that is really addressed the business value, by contrast with the work done with the Thales Digital Factory, which was much more technical.

3.3   Final Report Generation

All the posters presented above have been design under PowerPoint using a set a fonts that allow for acceptable legibility during a workshop, when printed in A0 format, and acceptable legibility for individual reading, in A4 format. Thus, when all the posters resulting from the workshops have been edited electronically, it is possible to generate a final cybersecurity report by printing the PowerPoint file. Our workshop participants found it easy to review the documents because the data was presented in the report in exactly the same way as it was during the workshop.

In addition to the posters themselves, we have added a cover page, and a conclusion page to the PowerPoint file. The conclusion page is the sole *text-only* page of the document: it summarizes the results and recalls the names of the workshop participants, their affiliation and role in the study. As a reference, the report generated for our SCADA IoT case-study is 16 pages in length, cover and conclusion pages included.

4   Conclusion

Lean management is trendy. This also concerns risk management, in particular in France, with the recent publication of the EBIOS-Risk Manager method by the French National Agency for Cybersecurity. However, if the new method fosters an agile approach of risk management, it does not provide the tools to support the mandated brainstorming workshops. In this paper we have proposed an innovative set of A0 posters to support the collection of risk management information during brainstorming workshops. By using these posters on a Thales internal cybersecurity course and on two real business case-studies, we have developed the optimal number and the content of each poster, bringing them to a level of maturity that is compliant with operational business cases. We have noticed during those case-studies that risk management using this technique is fun. It is a way of demystifying risk management, making it easier to understand, whilst remaining highly time-efficient. This format is especially appropriate during bid activities, or project kick-off. It also fosters a collaborative state of mind, recalling that system architecture securing is not the sole business of cybersecurity experts, but the result of a collaborative work involving the management, domain experts, the CISO and CIO.

On the posters, the allocated space for an activity, the scale axes, the positions and colours of sticky notes have all been fine-tuned to allow for the efficient capture of information, and also convey some subconscious messages and links between different bits of data. For example, if there is no more space to put your sticky note, then it is maybe time to move forward with the next activity, leaving whatever you still wanted to add for the next iteration.

We are currently promoting this set of posters for use in Thales by our Business Units as a possible tooling of the EBIOS-Risk Manager method. Our aim is to use this set of posters to support at least a first round of risk identification, assessment and treatment. Our case-studies show that initial results can be obtained after only 3–6 workshops of 2 h each. Iterations can then be run anew to go in more depth.

Beyond Thales internal uses, our poster templates will be made available to anyone requesting them under a Creative Commons CC BY-NC-SA licence.

We are fully aware that this approach will probably not scale to very large studies, on which the completeness and consistency of the risk management data need to be checked by electronic means. Beyond some 15 posters, the back office works becomes unacceptable. However, even on large studies, we believe that it is possible to start the study using this poster-based approach, to gain a momentum and community adherence to the risk management process, and then shift to some EBIOS compatible electronic software tool, like for example the RiskOversee products [10], to document the complete architecture, specific vulnerabilities and detailed attack scenarios.

As way forward, we are also considering the porting of this approach on a visual management tool, like iObeya© [11] or Framemo [12]. This improvement should allow for electronic sticky note support, decreasing the amount of back office work, and proposing a dynamic way of adjusting the layout regarding the workshop context. It should even be possible to use dedicated tactile screens to improve the user experience. An electronic visual management tool would also bring the possibility to perform this workshop remotely. However, we will be careful that electronic tooling does not override the intellectual approach and the dynamics of human collaboration. The licencing costs may also be an issue with some electronic sticky notes commercial frameworks.

We plan to continue to mature our approach on additional Thales internal case-studies, and on the upcoming H2020 Foresight research project (due to start in September 2019).

Acknowledgements. This research was partially funded by the French DGA CoSS-2 RAPID project. We wish to thank Sébastien Lhuillier and Thomas Baudillon from Thales SIX GTS FRANCE for their lead in the SCADA IoT case-study and Christophe Alix and Hélène Bachatène for their lead in the VLLAM/UTM case-study. Many thanks also to Fabien Caparros of ANSSI for his review of the poster templates, and to all the Thales engineers who made constructive comments on this work.

Appendixes

These appendixes provide the template of some posters which were only rapidly discussed in the main part of this paper, as well as multiple examples of filled in posters from two case-studies. To allow for a better understanding of the posters, we first provide an overview of the two case-studies.

Overview of the Case-Studies

The A0 posters were designed and validated using a running example of a Thales internal cybersecurity course and two real business case-studies. These case-studies are presented below.

IFE case-study. The In-Flight Entertainment (IFE) case-study was constructed for a Thales Learning Hub (TLH) adult professional training course on cybersecurity. Therefore, by contrast to the other two case-studies, the IFE case-study does not claim to provide a comprehensive framing of the system. This case is a toy case built to support educational goals.

According to the specifications given in the cybersecurity course, the IFE must provide free games, music and films to the airline passengers. "Free" means that there is neither credit card nor financial issues. The scope limited to games, music and films means that there are no connections with the avionics (and thereof limited safety-related issues), and no connections to the internet. It is therefore a very basic IFE. There is however a performance requirement: IFE availability should be above 99%.

VLLAM/UTM case-study. The poster templates presented in this paper were used to run a Very Low Level Airspace Management (VLLAM) and Unmanned Traffic Management (UTM) case-study. This case-study, run during the first semester of 2018, was very useful to raise the maturity of the workshop n°1 to workshop n°3 poster templates. Indeed, the workshops 1 to 3 were run twice: once for the overall system of systems, and once for the geofencing capability. The outputs of this case-study are not shown in this paper.

SCADA IoT case-study. Thales Ground Transportation intends to introduce Internet of Things (IoT) devices in its Metro Supervisory Control And Data Acquisition (SCADA) system. This obviously raises some questions about IoT cybersecurity. This case-study, run during the second semester of 2018, was very useful to raise the maturity of the workshop n°4 and workshop n°5 poster templates. The results of this case-study are extensively showed in the appendixes of this paper.

A Bit More About Logistics

As mentioned in the core part of this paper, EBIOS brainstorming sessions typically involve between 2 and 4 persons, in addition or including the facilitator. When paper versions of posters are used, this means that the sessions can (and should) be organised in relatively small rooms, with at least one wall where posters can be hanged, at a reasonable distance (i.e. 2 to 4 m) from the workshop participants.

Before each session, the facilitator should hang the set of A0 posters required for the session (see next sections for details), and distribute to each participant: (i) a single A5 cheat-sheet that recalls the workshop objectives and provides some hints as to how the session is going to be run; some cheat-sheets also provide tips, or knowledge bases, such as a list of classical threat sources; (ii) a set of colour sticky notes; the choice of colours depends on the workshop, as explained in the core part of this paper; and (iii) a felt pen; we recommend a different colour felt pen per participant, so that it authenticates the contributor; we also recommend a medium-sized felt pen tip, so that the writing remains readable up to a distance of 4 m.

Complementary Poster Templates in Support of Workshop n°1

A feared event is the negation of a security need on a primary asset. The proposed poster template allows for the study of up to ten feared events, represented by the pink sticky notes. Below the feared events, the impacts of the event may be listed, using orange sticky notes, against a severity scale represented by a vertical arrow, spanning from low severity (in green) to high severity (in red). As for the security need scale on the first poster (cf. Fig.2), it is important to capture here the relative severity of the impacts, rather than their absolute severity value. Figure 14 shows an extract of this poster filled in for the IFE case-study.

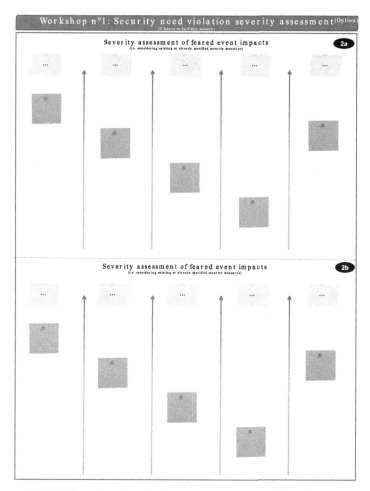

Fig. 10. Poster template for the severity assessment of feared events

The poster template in Fig. 11 supports the assessment of the implementation status of security controls. The poster is split in three sections corresponding to fully implemented controls, partially implemented controls and controls not implemented (from bottom-up).

In addition, the poster is split in three columns to allow for the classification of security controls. Thus, one would need three poster instances to cover the 7 families of functional requirements defined in IEC 62443-3-3 [13] ... but we do not recommend dealing with too many details with the poster-based approach. An exhaustive analysis of hundreds of security controls will best be managed using some dedicated electronic tool.

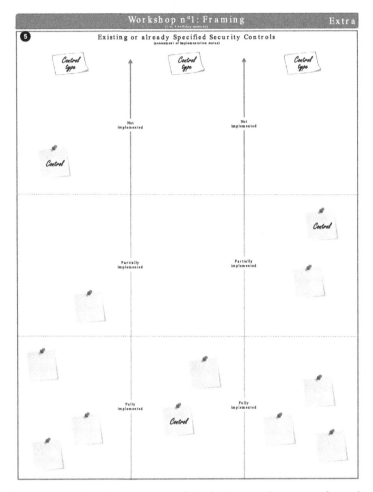

Fig. 11. Poster template for the assessment of the implementation status of security controls (Color figure online)

Examples of Posters Produced During Workshop n°1

Figure 12 shows the first poster supporting workshop n°1 as it was filled in for the SCADA IoT case-study. To begin with, the study objectives were capture: (i) Identify and manage IoT-related risks, in particular to help design the future IoT devices in a secure manner; (ii) Establish a risk assessment baseline; and (iii) Establish an IoT migration strategy.

On this poster, we can see here that quite a large number of stakeholders have been captured, including regulatory bodies and public services.

In terms of primary assets, we can clearly see two large groups: on the right side of the security need axis, a group of assets with very strong integrity and availability

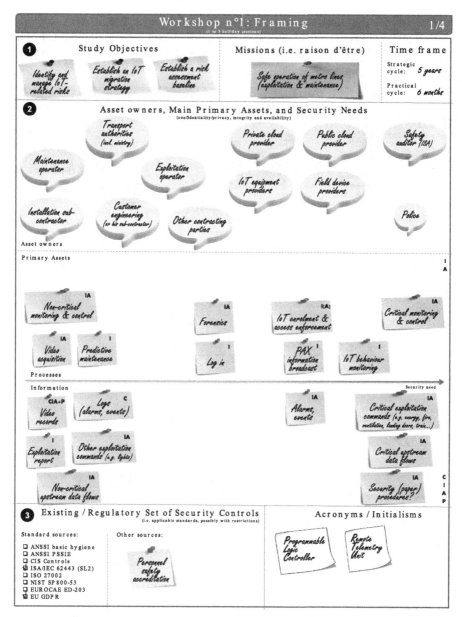

Fig. 12. First poster supporting workshop n°1, applied to a SCADA IoT case-study (Color figure online)

needs to ensure the core mission of the system; on the left side, non-critical services and data. On the latter, the integrity and availability needs remain predominant, but some confidentiality and privacy needs also appear.

Upon starting the study, we were given one important input. The SCADA currently complies and should continue to comply with Security Level 2 (SL2) of the IEC 62443 standard [14]. This regulatory constraint was registered at the bottom left of the poster.

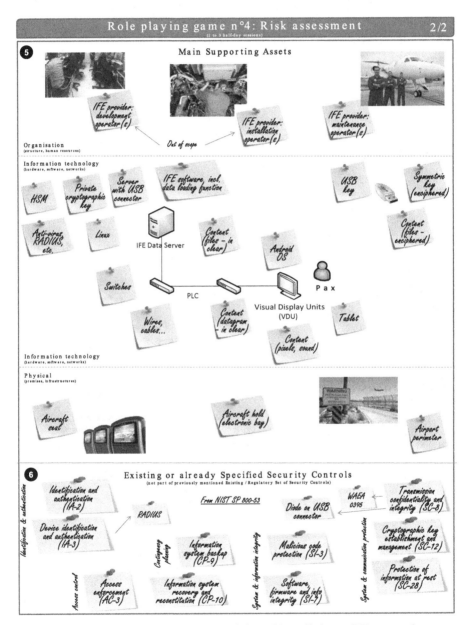

Fig. 13. Second poster supporting workshop n°1, applied to an IFE case-study

Figure 13 presents an example of the second poster supporting workshop n° 1, applied this time to the IFE case-study. The IFE case-study is focused on the operational exploitation of the IFE, i.e. it excludes the development and installation parts of the system's lifecycle. Thus, we have only identified the IFE maintenance operator(s) as organisational supporting assets. For IT supporting assets the list is more extensive. To keep the story short, let us just focus here on the supporting assets of the copyrighted content, i.e. films, music and games. In the IFE case-study, copyrighted content is a primary asset with obvious confidentiality needs. The copyrighted content exists physically in many forms, e.g.: (i) as an enciphered file in the USB stick that the IFE maintenance operator carries to perform content update; (ii) as a file in clear on the disk of the IFE data server; (iii) as a datagram in the cables and switches between the IFE data server and the Visual Display Units (VDUs); and (iv) as pixels and sound on the VDUs. Last but not least, we have identified three physical supporting assets: the aircraft seat, the electronic bay in the aircraft hold, and the airport perimeter.

The fifth and last section of the poster relates to existing or already specified security controls. In the IFE case-study, the NIST SP 800-53 standard [15] has been used, therefore the poster shows the NIST identifiers of the security controls, e.g. Device Identification and Authentication (IA-3), and Transmission confidentiality and integrity (SC-8).

For some controls, the poster also shows how these controls are/will be implemented. For example, IA-3 will be ensured using a classical Remote Authentication Dial-In User Service (RADIUS), whilst Transmission Confidentiality and Integrity (SC-8) will be ensured using the (now obsolete) domain-specific WAEA 0395 standard [16], and a diode on the USB connector of the IFE server.

The identification of supporting assets is important to later identify how the security needs of the primary assets may be breached. Each supporting asset has its vulnerabilities and may be attacked in its own way. If we stick to the example of the copyrighted content it is possible to: (i) mug the maintenance operator and snatch the USB key with the enciphered files of the copyrighted content during a maintenance operation; (ii) to steal the removable disk in the IFE data server with the copyrighted content in clear; (iii) to sniff the network; or (iv) film the VDU with a smart phone, whilst recording the sound via the jack plug.

The existing or already specified security controls are also important to later assess the likelihood of the aforementioned security breaches. For example, the existence of a RADIUS makes it a bit less likely for an attacker to spoof a VDU with its own recording device.

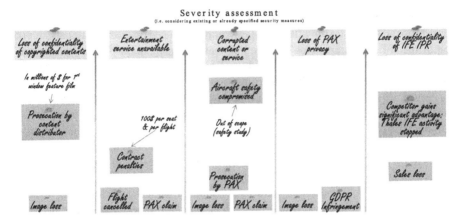

Fig. 14. Severity assessment of the IFE feared events (Color figure online)

Figure 14 shows an extract of the workshop n°1 optional poster, filled-in for the IFE case-study. This cropped poster illustrates five feared events. For example, if we consider that copyrighted content needs to be confidential in the IFE case-study, then "loss of the confidentiality of the copyrighted content" or "a first-window feature films becomes freely accessible on the Internet" make two perfect examples of feared events. It can be seen here that the most severe feared event impacts relate to the violation of the confidentiality of the copyrighted content (1st column) and to the violation of the confidentiality of the IFE developer's know-how (5th column). This is because the corruption of the content or the service leading to a safety event, e.g. the display of a malicious message such as "All passengers, please move to the rear of the aircraft" with its probable dramatic effect on the aircraft balance, is considered out of scope of this study, as it is normally already covered by the safety case. Following this work, the IFE system has been categorized as "Moderate" according to the NIST SP 800-53 standard [15]. This categorisation pulls a significant set of security controls, some of which are listed on the poster discussed above (cf. Fig. 13).

Complementary Poster Template in Support of Workshop n°2

Figure 15 shows the extra poster for workshop n°2. This poster extends the main workshop n°2 poster (cf. Fig. 4) to cope with additional risk sources. This poster may be used as often as required, however, up to now, in all our case-studies, a single instance of this extra poster proved sufficient.

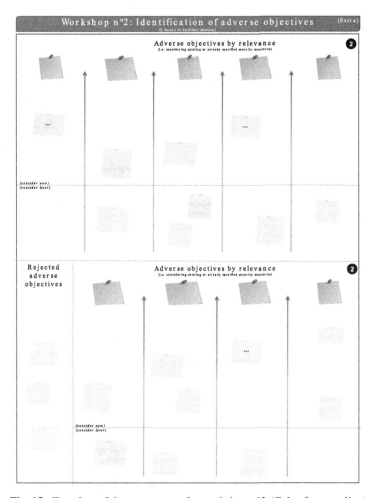

Fig. 15. Template of the extra poster for workshop n°2 (Color figure online)

Example of Poster Produced During Workshop n°2

Figure 16 shows the main poster for workshop n°2 filled in for the SCADA IoT case-study. It can be seen here that the most relevant risk sources are malevolent ones, essentially external, but the internal rogue employee is also highly considered. In the lower part of the poster, the objectives of the latter even appear as the most relevant adverse objective, i.e. Vengeance through sabotage and/or Denial of Service (DoS), and the self-creation of maintenance workload.

The poster also shows that two risk sources have been rejected: the competitor because he would have easier ways of acting, and the meteorological conditions, because of past experience with ruggedized equipment. As a consequence, the objectives of the competitor have been listed but obviously rejected.

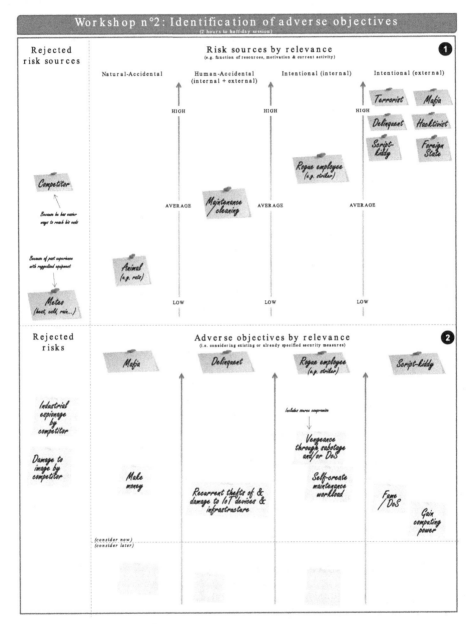

Fig. 16. Poster of workshop n°2 filled-in for the SCADA IoT case-study (Color figure online)

Complementary Poster Template in Support of Workshop n°3

Figure 17 proposes a poster template to register all risks, as identified during workshop n°3. The main objective of this register is to allow for the referencing of risks by their number, under the format Rxx, rather than pull the often long description of the scenarios.

In addition, the template allows capturing additional comments for each risk, which is something that was very much restricted with our poster-based approach up to now.

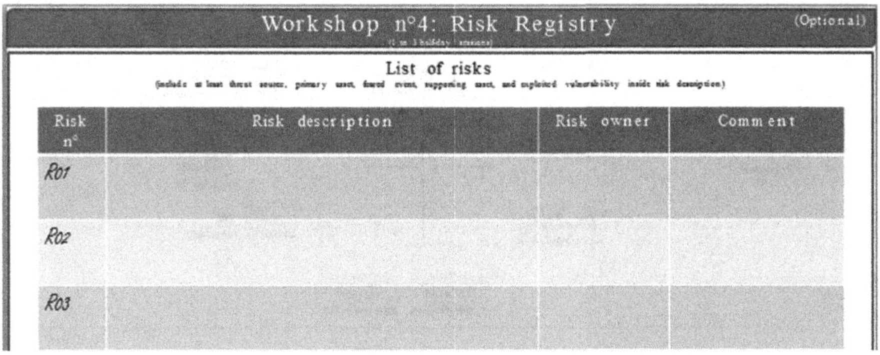

Fig. 17. Poster template to be used as risk register (cropped)

Example of Poster Produced During Workshop n°3

Figure 18 shows the first poster of workshop n°3 filled in for the SCADA IoT case-study. On the upper part of the poster, it can be seen that there are no critical stakeholders: when stakeholders have high privileges on the system, they are trustworthy, and if they are untrustworthy, then they do not have special privileges. Still, attention is called upon the local maintenance operators, third-party suppliers, direct sub-contractors, physical security services and company executives. The latter was commented as being particularly difficult to cope with, as it is often difficult to refuse access rights to one's boss, even if he does not need it.

The lower part of the poster is used to capture how the risk sources will proceed to reach their objective(s), at a very gross-grain level, e.g. in the SCADA IoT case-study, "Mafia installs ransomware on an IoT device connected to the trusted network using spearfishing targeted at the maintenance operator". It can be seen here that there are quite a few risks with a pretty high severity. Risks that relate to both integrity and availability are located astride the severity axis.

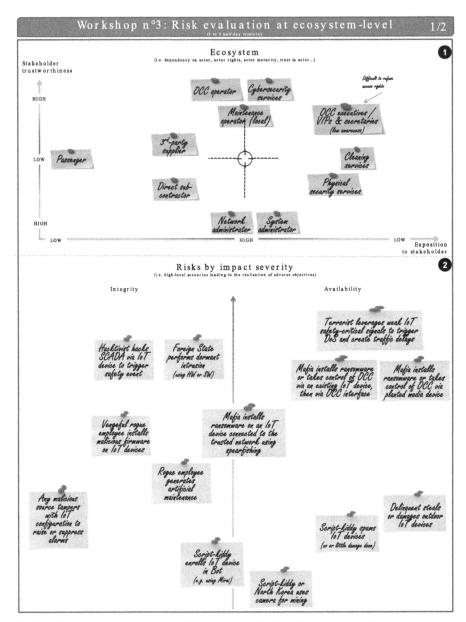

Fig. 18. First poster of workshop n°3 filled-in for the SCADA IoT case-study (Color figure online)

It should be reminded here that ultimately the criticality of a risk depends on both the severity of its impacts and the likelihood of its occurrence. At this stage the likelihood has not yet been studied.

Example of Poster Produced During Workshop n°4

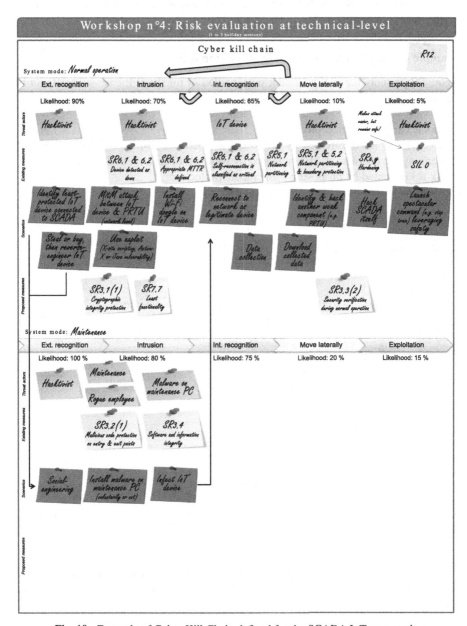

Fig. 19. Example of Cyber Kill Chain defined for the SCADA IoT case-study

Figure 19 shows a Cyber Kill Chain created for our SCADA IoT case-study. The upper part of the poster shows a full blown scenario, in which the attacker has significant cybersecurity knowledge to reverse-engineer the IoT device and exploit some communication vulnerability. The lower part of the poster shows an alternative path, in which social engineering is used to benefit from the collusion of a rogue employee. The social engineering path has been assessed as more likely.

Example of Poster Produced During Workshop n°5

Figure 20 presents the risk treatment poster filled in for the IFE case-study. It can be seen here that:

- risks R5 and R6 are accepted;
- risks R1 and R2 are treated by proposing the perform some media control and by changing the obsolete WAEA standard by its newest edition, i.e. WAEA 0403;
- risk R3 is treated by proposing some media protection;
- risk R4 is shared by contracting an insurance.

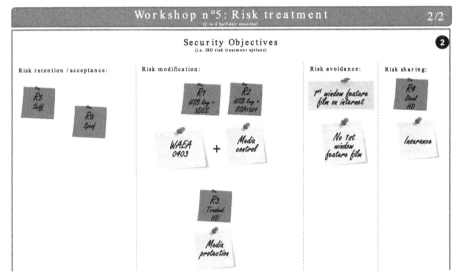

Fig. 20. Example of risk treatment defined for the SCADA IoT case-study (cropped) (Color figure online)

We can also see that the event of finding a first-window feature film on internet can be avoided by not showing any first-window feature film. Here, the airline may not agree with the treatment option and plan.

References

1. ANSSI. EBIOS Risk Manager, version 1.0 (in French). Agence Nationale de la Sécurité des Systèmes d'Information, Paris (2018)

2. ISO/IEC 27005. Information technology—Security techniques—Information security risk management. International Organization for Standardization/International Electrotechnical Commission, Geneva (2018)
3. NIST SP800-64r2. Security Considerations in the System Development Life Cycle. National Institute of Standards and Technology, Gaithersburg (2008)
4. 47-DDQ-GRP-EN. Cybersecurity Engineering Guide (commercial-in-confidence). Thales Chorus 2.0. Accessed 31 Jan 2018
5. CESG. HMG IA Standard Numbers 1 & 2, Information Risk Management. National Technical Authority for Information Assurance, Cheltenham, Gloucestershire, UK (2012)
6. CESG. HMG IA Standard Numbers 1 & 2 – Supplement, Technical Risk Assessment and Risk Treatment. National Technical Authority for Information Assurance, Cheltenham, Gloucestershire, UK (2012)
7. Lockheed Martin. The Cyber Kill Chain. https://www.lockheedmartin.com/en-us/capabilities/cyber/cyber-kill-chain.html. Accessed 6 March 2019
8. OMG. Software & Systems Process Engineering Metamodel (SPEM). Object Management Group (2008)
9. ANSSI. EBIOS Risk Manager Accreditation: Tools to support Cybersecurity Risk Management (in French only) (2019). https://www.ssi.gouv.fr/administration/management-du-risque/la-methode-ebios-risk-manager/label-ebios-risk-manager-des-outils-pour-faciliter-le-management-du-risque-numerique/. Accessed 16 May 2019
10. RiskOversee. Tool-up your EBIOS analysis. ALL4TEC (2019). https://www.riskoversee.com/en/cyber-architect-en/. Accessed 16 May 2019
11. KapIt. Digital Visual Management for Lean & Agile companies. https://www.iobeya.com/. Accessed 22 May 2019
12. Framasoft. Framemo. https://framemo.org/demo. Accessed 22 May 2019
13. IEC 62443-3-3. Industrial communication networks - Network and system security - Part 3–3: System security requirements and security levels. International Electrotechnical Commission (2013)
14. IEC 62443. Industrial communication networks - Network and system security. Industrial Automation and Control System Security Committee of the International Society for Automation (ISA)
15. NIST SP 800-53. Security and Privacy Controls for Federal Information Systems Federal Information Systems, Special Publication 800-53, Revision 4. National Institute of Standards and Technology, Gaithersburg (2013)
16. DMD-TC. WAEA Specification 0395, Content Delivery for In-Flight Entertainment. Digital Media Distribution Technical Committee of the World Airline Entertainment Association Technology Committee (WAEA-TC), Virginia, USA (2001)

Conceptual Abstraction of Attack Graphs - A Use Case of securiCAD

Xinyue Mao[1], Mathias Ekstedt[1][(✉)], Engla Ling[1], Erik Ringdahl[2], and Robert Lagerström[1]

[1] KTH Royal Institute of Technology, Stockholm, Sweden
{xinyuem,mekstedt,englal,robertl}@kth.se
[2] Foreseeti AB, Stockholm, Sweden
erik.ringdahl@foreseeti.com

Abstract. Attack graphs quickly become large and challenging to understand and overview. As a means to ease this burden this paper presents an approach to introduce conceptual hierarchies of attack graphs. In this approach several attack steps are aggregated into abstract attack steps that can be given more comprehensive names. With such abstract attack graphs, it is possible to drill down, in several steps, to gain more granularity, and to move back up. The approach has been applied to the attack graphs generated by the cyber threat modeling tool securiCAD.

Keywords: Attack graph · Conceptual modeling · Cognitive simplification · securiCAD

1 Introduction

The complexity and size of IT systems are growing and as a result so are the attack graphs that represents possible attacks against them. It is important to make sure that the attack graphs are useful and easy to interpret even as they grow. This short paper presents a solution to the problem of visualizing large attack graphs by using abstractions.

This work was driven by a need to simplify the attack graphs generated in the attack simulation tool securiCAD [2]. In securiCAD attack graphs are generated using a fixed attack step library and graph generation logic encoded in a domain specific language. The visualization of the generated graphs follows that same terminology of the semantic level of the library. As this language is quite extensive the generated attack graphs quickly become complex and difficult to grasp, as seen for example in Fig. 1. This paper describes a solution to generate abstracted visualizations of attack graphs with two objectives: (1) the abstraction should be formally sound and reversible, and (2) the abstractions should be

This work has received funding from the Swedish Civil Contingencies Agency through the research centre Resilient Information and Control Systems (RICS) as well as the SweGRIDS competence center.

© Springer Nature Switzerland AG 2019
M. Albanese et al. (Eds.): GraMSec 2019, LNCS 11720, pp. 186–202, 2019.
https://doi.org/10.1007/978-3-030-36537-0_9

understandable and make sense to the users of the attack graph. The the paper contributes both with an approach to form visually simplified attack graphs particularly from asset-based attack graph formalisms, as well as a case study on the securiCAD tool where the approach was applied. The approach is limited to simplifying already generated attack graphs. Consequently, the question of the correctness of these attack graphs or the computational challenges of computing them is outside the scope of this paper.

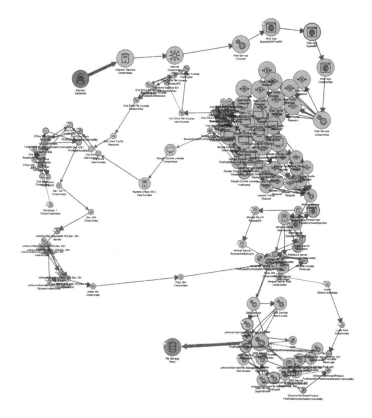

Fig. 1. An example of a visually complex attack graph in securiCAD. (Attacker starting point - red circle, attacker goal - blue circle). (Color figure online)

2   Related Work

The problem of making information more accessible in complicated and large attack graphs can be solved by several different methods as outlined in the attack graph taxonomy by Kaynar [8]. One method of simplifying is, for example, hierarchical division. In our solution the hierarchical division is derived from the securiCAD syntax. In the tool CyGraph [14], which uses multiple data sources to construct an attack graph, the different data sources are used as layers. The layers are, for instance, the network infrastructure layer and the cyber threats layer. Other research has based the division on attribute values or the connectedness of the attack graph [15]. Another example of division is HARM, a hierarchical attack representation model [4]. HARM divides the graph in two hierarchical levels, the network and the vulnerability levels.

An alternative solution to simplify attack graphs is to aggregate parts of the graph. The aggregation can be achieved by grouping the attack steps or network objects [9]. However, this work was presented as a suggested approach of reducing the complexity but did not present a method. The approach was also intended to reduce the complexity of generating graphs, rather than presenting them. Please note that, in this paper, we do not aim to reduce the complexity of graph generation. This paper is concerned with simplifying attack graphs' representation. Homer et al. presented another approach to aggregation where they remove attack steps that they consider "useless" [3]. Their definition of "useless" is that the attack step is not necessary to understand the security vulnerability.

The Network Security Planning Architecture (NetSPA) uses a method of pruning to simplify the representation of an attack graph [1]. The user can prune the attack graph by choosing a specific goal state and only visualize the attack steps that ends with that specific state. This method of pruning an attack graph has not been implemented in the solution presented in this paper.

A common denominator for some of the identified related work is that the aggregation is performed according to fixed rules built upon the specific attack graph elements. In our work we present a solution for building aggregation patterns (that can be changed over time) as well as a specific set of patterns for the securiCAD tool.

The patterns that are constructed in our work are named according to a visualization vocabulary. The intention of these patterns is to align with established terminology. Examples of such resources are MITRE's Common Attack Pattern Enumeration and Classification (CAPEC) [13] and ATT&CK matrix [12]. However, in our work a bespoke vocabulary was developed matching the securiCAD tool.

There are methods of reducing the complexity of graphs' visual representations in ways not used in the solution presented in this paper. By working with for example different thickness or colors of lines, it is possible to add more information to the graph without adding more items [10].

Finally, there are methods for reducing attack graph complexity for other reasons than visualization. These reasons can be, for instance, to reduce computational complexity as seen in the survey by Hong et al. [5]. However, these related works are not included because they fall out of scope of this paper.

3   securiLang

The securiCAD tool [2] is generating its attack graphs according to the logic of a domain specific language called securiLang. In brief the securiLang consists of Assets (e.g. Network, Host, Service, Data flow) that can have Associations to each other (e.g. a Host can either Root execute or User execute a Service). To Assets there are Attack steps associated (e.g. a Data flow can be Eavesdropped, Replayed, or DoSed). In addition, Assets also have Defenses associated. However, this paper is only looking at attack graphs, not the full defense graphs, so they are not further discussed here. The full list of assets and their associated attack steps are presented in Appendix A.

Furthermore, securiLang contains rules for how attack graphs are generated when instance models are built following the language. Attack steps have potential parent and child steps, depending on how the instance models is constructed. E.g. Figure 2 demonstrates the relation between asset Access Control and its attack step AccessControl.Access. A potential parent to this step is Host.UserAccess, however this is only true if the Access Control asset is having an Authorization association to the Host asset. If the Access Control Authorizes a Service instead it will be the Service.Connect that leads to AccessControl.Access, and so on. In this example AccessControl.Access only has child attack steps located at the Access Control asset.

Fig. 2. An example of Access Control showing the association of an attack step

The full logic in terms of association traversal and OR/AND attack step dependencies are omitted here to avoid complexity unnecessary for the purpose of the paper. Furthermore, the securiCAD tool is conducting probabilistic calculations of the attack graphs and aggregates a time-to-compromise value over all possible attack paths enabled by an instance model which quickly leads to a myriad of attack vectors. However, these dimensions of the attack graph generation and calculation is not discussed here since the attack graph aggregation suggested in this paper follows the same structure for all branches of an attack graph. In the next chapter we will go into the abstraction mechanism.

4   Attack Graph Abstraction

Our abstraction mechanism addresses attack graph languages based on relational models following our previous work such as [6,7] and [16]. The aggregation approach is illustrated in the conceptual model in Fig. 3.

The original attack graph language is depicted in the yellow classes. We consider this the level 0. From the meta model of this language we assume three things; that it includes an `Attack Step` class, an `Asset` class, and an `Association` relating the two classes. This is true for our previous work, but the intention is to leave as few requirements on the original language as possible[1]. Not strictly needed to apply the approach, but later used in our use case for convenience, the Figure also illustrates that `Assets` can have `relationships` between each other. In the original language these relationships are however instantiated so that e.g. a *Host* `Asset` *executes* a *Client* `Asset`.

On top of this original language we introduce two new classes; the `Abstract Attack Step` and the `Abstract Asset`. The purpose of the former class is to enable aggregation of `Attack Steps` and `Abstract Attack Steps` in leveled hierarchies. Thus we also introduce an `Aggregation` relationship between `Abstract Attack Steps` and `Attack Steps` as well as a self reference for `Abstract Attack Steps`. Here we follow the standard semantics of aggregation, for instance used in UML, where the aggregator is nothing more than, and cannot exist without, its constituents. More precisely, we introduce two types of aggregations (not illustrated separately in the Figure to avoid clutter); `Mandatory Aggregation` and `Optional Aggregation` since not always are all possible sub steps present in the illustrated attack graph[2].

In order to create the aggregation hierarchies, we introduce a `Level` variable. We stipulate that the original language is level 0 and then for every introduced aggregation the `Level` is increased. However, we do not restrict aggregation so that all constituents need to be of the same `Level` and consequently the `Level` value has to be determined by the highest constituent `Level`. The aggregator `Level` must be increased with at least 1 but we also allow to add any (natural) number in order for the aggregation designer to build "conceptually even" layers. Similarly to the original language we also stipulate that `Abstract Attack Steps` must have an `Association` to an `Asset` or an `Abstract Asset`. With the `Abstract Asset` enable the possibility to associate `Abstract Attack Steps` with any type of asset that is deemed useful. And just as we assume that `Attack Steps` and `Assets` come with `Names` (or identifiers) from the original language, also the `Abstract Attack Steps` and `Abstract Assets` are given `Names`.

Given these rules for abstracting attack graphs there are many patterns imaginable for how this can be done. In the case study presented in this paper mainly two patterns have been used. Firstly, aggregation of a fixed series of parent/child

---

[1] Obviously the exact naming of the classes and their relationship is not relevant.

[2] This could be due to the attack steps being aggregated by an OR attack step, but it could also be that not all attack vectors are displayed at the same time. In securiCAD for instance only highly probable attack vectors are illustrated.

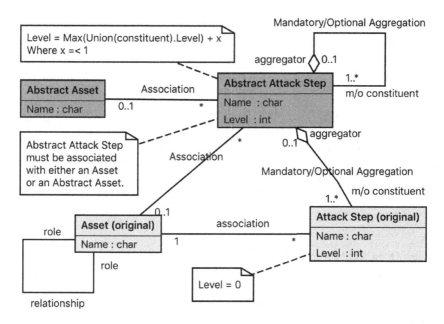

Fig. 3. Attack graph abstraction overview

attack steps following on each other according to the logic of the original language. An example of such a case is the one illustrated in in Fig. 2. Secondly, we have based aggregations on the `Asset` they are `Associated` with. An `Abstract Attack Step` is defined that represents a full compromise of the `Asset`. Here we want to aggregate all `(Abstract) Attack Steps` that relate to this particular `Asset`. This pattern allows for conceptually organizing the `Assets` into the layers, so that one `Asset`'s full compromise `Abstract Attack Step` is then aggregated into another `Asset`'s full compromise `Abstract Attack Step`.

5 Abstracting securiLang

Above we have presented the general mechanisms devised for abstracting model based attack graphs. We will now move on to describing a suggested abstraction for securiLang[3]. In this work we have devised six abstraction levels, three attack step based abstractions(Level 1–3) and three asset based abstractions(Level 4–6).

For each level of abstraction, we merge the original attack steps into one abstracted step according to the patterns introduced in the last section. All securiLang abstractions made are found in Appendix B. To illustrate the work we take `Exploit vulnerabilities and compromise Client` as an example to

---

[3] The work is based on the securiLang version contained in securiCAD v1.4.

explain how the abstracting operation works. Table 1 contains a snippet from the full table in Appendix B.

Table 1. Abstraction of attack steps in asset client

Original attack steps	1st level	2nd level	3rd level
Client. FindExploit	O	Find exploit	Exploit vulnerabilities and compromise Client
Client. DeployExploit	O	Deploy exploit	
Client. BypassIDS			
Client. BypassAntiMalware			
Client. Compromise	O	Compromise Client	

The 'O' in Table 1 means that in this level, all the abstracted steps remain the same with former level. For example, the 'O' in the first line of Fig. 1 means the same with the name in former level, in this case, this 'O' equals to Client.FindExploit. At the second level, the single steps are translated to Find exploit and Compromise client, while the three steps in the middle are grouped into Deploy exploit. Then in the third level, the abstracted steps in the second level can be merged into Exploit vulnerabilities and compromise Client. Figure 4 illustrates the abstraction in Table 1 graphically.

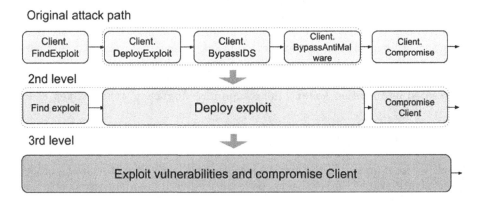

Fig. 4. Abstracted attack step of Access Control from original attack path to Level 3

Semantically Find exploit denotes that attacker is able to find an exploit to use on a Client. Deploy exploit means that the deployment of found and developed exploits for the Client's Software Product. After a successfully executed exploit is deployed, bypassing some protection mechanisms, the Client is Compromised. When making the higher level abstractions we want the terminology to be as intuitive and self explaining as possible, in this case exploit can be

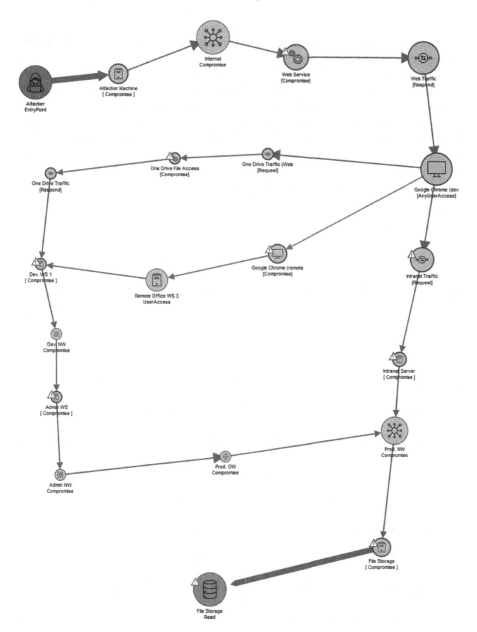

Fig. 5. An example of an abstracted attack graph in securiCAD on Level 4

explained as taking advantages of a bug or a vulnerability, which is not captured in the original securiLang terminology. Thus, in this case level 3 abstract attack step name was chosen as **Exploit vulnerabilities and compromise Client.** On level 2 we find examples of the use of the same name as used in the Level

0 language, and in two instances the semantics is kept, while in one case the semantics is changed (expanded so that also defense bypassing is included in the `Deploy exploit` abstract attack step).

Altogether we have done four asset based abstractions; `Client Compromise` and `Service Compromise` are found on Level 4, `Host Compromise` on Level 5, and `Network Compromise` is on the 6th and final Level. In securiLang these four assets; `Client`, `Service`, `Host`, and `Network`, each attack steps called `Compromise`, thus it was straight forward to develop new new `Abstract Attack Step` with the same name as the original language new `Attack Step` to create these levels.

As an illustration of the end result of this work it has enabled us to simplify the visually complex attack graph in Fig. 1 to an attack graph as seen in Fig. 5, on Level 4.

6   Evaluating the securiLang Abstraction

In order to evaluate our design of the securiLang abstraction we were addressing the following quality criteria (relating to our second objective in the Introduction):

1. The vocabulary chosen as representation provides associations that accurately reflects the attack steps on level 0.
2. Each of the representations are made so that the levels feels conceptually even in terms of level of perceived detail.
3. The design of the six levels are perceived as intuitive.

The evaluation was made in two phases; firstly two senior security experts that have previously worked with securiCAD (but not involved in the abstraction project) were interviewed and secondly a survey was sent out to cyber security students that had not encountered the tool.

6.1   The Interviews

For the interviews, we used a questionnaire where the respondents were asked to rate statements and answer questions. To address the first evaluation criterion statements such as *"Root login to Host* well describes the original attack process?" were posed with answering options on a five-level scale from *Strongly disagree* to *Strongly agree*. For the second criterion, questions such as "Do you agree that *Access* and *Root login* can be merged into one attack step in the 1st level?" were posed with the same answering option. Both statements were followed by an open question to allow the interviewee to comment freely on their answer.

In general, the two respondents gave the original design a good evaluation. However, several smaller changes were suggested and the model presented in this paper is the refined version after this input. For instance, `Use legitimate access` used to be called `Extract password from user`, which represented the two attack steps `ExtractFromUser` and *Compromise*. Another example is *FindExploit* and `DeployExploit` that were originally grouped together, while `BypassIDS`, `BypassAntiMalware` and `Compromise` were aggregated. The respondents instead suggested that `DeployExploit`, `BypassIDS` and `BypassAntiMalware` are grouped into one abstracted step instead.

6.2 The Survey

An online survey was created where participants rated the attack steps representations from strongly disagree to strongly agree. Altogether 24 students with major in information technology and at least basic skills in cyber security participated.

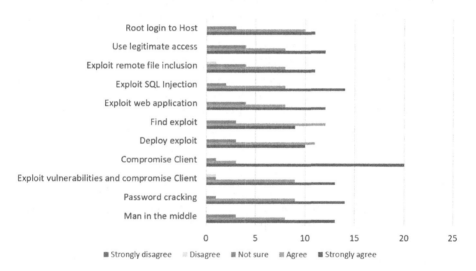

Fig. 6. Results of survey of single abstracted representation with corresponding numbers

Related to our first and third criteria, here the participants rated eleven selected attack step abstractions based on their intuitive perception and understanding, ranging from strongly disagree to strongly agree. Figure 6 shows the result of the survey for the different attack step abstractions. All representations except for `Find exploit` and `Deploy exploit` were rated with *Strongly agree* as the most popular choice. Overall only few *disagreed* and no participant chose *Strongly disagree*. When observing the results for *Agree* and *Not sure*, the number of people in these categories are evenly distributed among all the representations. Thus, it is concluded that each representation gets almost the consistent recognized level. For all representations, approximately 90% of the participants have chosen agree or strongly agree. This shows that most people agree with attack steps representations, believing they can well describe the original attack process and provide an easily understandable interactive view, which could meet the first and third criterion.

7   Future Work

This project reports on early and exploratory work. Consequently it is natural that several avenues for future work has been identified from this project. From a practical point of view making the abstraction a fully maintainable and configurable capability is the most important for further adoption of the idea. First of all, securiLang (and any other asset based attack graph language) evolves over time with new types of attacks and assets, so likely the attack aggregations also need to be updated. Moreover, securiCAD supports several different languages and each would need to have its language specific attack aggregations.

We can also envision that one would like to abstract the same attack graph according to different patterns depending on who is looking at it and for what purposes. In the case study presented here the aggregation was driven mainly by the experiences at the tool vendor of some attack path segments had been found difficult to explain to tool users in combination with some gut feeling of nice-to-have features. The aggregation could also be addressed e.g. from an alignment point of view where the aggregations are devised to map some more well established terminology such as Mitre's CAPEC [13] and ATT&CK [12], or Lockheed Martin's Kill Chain concept. Thus ongoing work is now to develop a configuration structure that enables to define the conceptual attack graph abstractions for any language in a format that is both readable by securiCAD and easy to expand and understand as a language designer. As of now, the configuration structure is in essence a single JSON file that for each level of abstraction defines a set of rules which are applied sequentially on the attack graph, such as the below code snippet.

```
Level 2: {
  Client: [
    {
      collect: [
        Client.DeployExploit,
        Client.BypassIDS,
        Client.BypassAntiMalware
      ],
      replacement: {
        class: Client,
        attackstep: Deploy exploit
      }
    }
  ]
}
```

From a more theoretical point of view, formalizing the aggregation approach suggested here into a (meta) language with formal semantics is a natural next step. Devising and structuring design patterns for aggregation can further be of interest. In addition, adding configuration features for syntax and form editing (coloring, sizing, etc.) is likely to improve attack graph readability significantly.

8   Conclusion

The purpose of this work was to address the problem of attack graphs that are difficult to understand as they grow large and complex. Specifically, the attack graphs generated by the tool securiCAD constituted a case study. An overall approach was suggested and applied in the case study. The results are generally positive. Firstly the attack graph abstraction approach was found viable and useful for conducting the case study. Secondly, the abstraction patterns developed in the case study were also promising. Even though the evaluation of the case study patterns was rather weak from a methodological standpoint at least all indications suggest that the patterns are fit for its intended purposes. Also the intuition at the case study company is that this is important correct enough to further develop the idea and implement support for it in the securiCAD tool. But to come to a more universal conclusion on the appropriateness of these particular aggregation patterns would obviously require more validation studies.

And for the generic conceptual abstraction approach per se has only been tested with a single case study. Obviously this is far from a full validation of the approach. Analytically we can however conclude that the approach is primarily suitable for use cases where the attack graphs we want to abstract follow predictable patterns since aggregation patterns need to be created manually.

Furthermore, the case study does not make use of `Abstract Assets` class. To use introduce for instance highly abstract assets as means for abstraction seems however reasonable. E.g. for large ICT infrastructures it could be interesting to. introduce an `Abstract Asset` called `Zone` onto which one or several `Network.compromise` could be aggregated into an `Abstract Attack Step` labelled `Zone.compromise`.

This paper is based on and extends the work presented in the Master thesis by Mao [11].

A   Appendix

securiLang Assets and Attack Steps

Assets	Attack steps
AccessControl	Access, ExtractPasswordRepository, NonRootLogin, RootLogin
Client	BypassAntiMalware, BypassIDS, Compromise,DenialOfService, DeployExploit, FindExploit, UserAccess
Dataflow	Access, DenialOfService, Eavesdrop, ManInTheMiddle, Replay, Request, Respond
Datastore	Delete, Read, Write
Firewall	Compromise, DiscoverEntrance
Host	ARPCachePoisoning, BypassAntiMalware, BypassIDS, Compromise, DenialOfService, DeployExploit, FindExploit, PhysicalAccess, PrivilegeEscalation, USBAccess, UserAccess
Keystore	Delete, Read
Network	ARPCachePoisoning, Compromise, DNSSpoof, DenialOfService
Router	Compromise, DenialOfService, Forwarding
Service	ApplicationLogin, BypassAntiMalware, BypassIDS,Compromise, Connect, DenialOfService, DeployExploit, FindExploit, NonRootShellLogin, RootShellLogin, UserAccess
SofwareProduct	DevelopExploitForPublicPatchableVulnerability, DevelopExploitForPublicUnpatchableVulnerability, DevelopZeroDay, FindExploitForPublicPatchableVulnerability,FindExploitForPublicUnpatchableVulnerability, FindPublicPatchableVulnerability, FindPublicUnpatchableVulnerability
UserAccount	Compromise, ExtractFromUser, GuessOffline, GuessOnline
WebApplication	BypassWAFViaCI, BypassWAFViaRFI, BypassWAFViaSQLInjection, BypassWAFViaXSS, DiscoverNewVulnerability, ExploitCommandInjection, ExploitRFI, ExploitSQLInjection, ExploitXS

securiLang is further described at https://community.securicad.com/securilang-reference-manual/

B   Appendix

Abstraction Patterns of securiLang

Asset	Attack Step	1st Level	2nd Level	3rd Level	4th Level	5th Level	6th Level
Access Control	Access UserAccount(root).Compromise RootLogin	Root login to Service / Host	×	×	×	×	×
	Access UserAccount(non-root).Compromise NonRootLogin	Non-root login to Service / Host	×	×	×	×	×
User Account	Access ExtractPasswordRepository UserAccount.GuessOffline UserAccount.Compromise	Password cracking	×	×	×	×	×
	ExtractFromUser Compromise	Use legitimate access	×	×	×	×	×
	AccessControl.Access GuessOnline Compromise UserAccess	Guess possible credentials	×	×	×	×	×
Client	SoftwareProduct.FindExploit / DevelopExploit / DevelopZeroDay / FindEndOfLifeVuln	Access to Client	×	Exploit vulnerabilities and compromise Client	Client Compromise		
	FindExploit	O	Find exploit				
	DeployExploit	O	Deploy exploit				
	BypassIDS						
	BypassAntiMalware	O					
	Compromise		Compromise Client				
	Host.DenialOfService DenialOfService Dataflow.DenialOfService Access	Denial of Service	×	×	×	×	×
Dataflow	Service.DenialOfService / Client.DenialOfService / Network.DenialOfService DenialOfService	Denial of Service	×	×	×	×	×
	Access Eavesdrop	Eavesdrop	×	×	×	×	×
	Access ManInTheMiddle Request / Respond	Man in the middle	×	×	×	×	×
	Access Replay Request	Replay	×	×	×	×	×
	Service.Connect Respond	Service.Connect	×	×	×	×	×
	Client.UserAccess	Client.UserAccess	×	×	×	×	×
Network	ARPCachePoisoning DNSSpoof	ARP Cache Poisoning DNS Spoof	× ×	× ×	× ×	× ×	Network Compromise
	Router.DenialOfService DenialOfService Dataflow.DenialOfService	Denial of Service	×	×	×	×	

* Steps in grey are optional in the aggregation

Asset	Attack Step	Attack Steps Based Abstraction			Assets Based Abstraction		
		1st Level	2nd Level	3rd Level	4th Level	5th Level	6th Level
Keystore	Delete / Datastore.Delete	Damage data	X	X	X	X	X
	Read / UserAccount.Compromise	Compromise UserAccount	X	X	X	X	X
	Read / Datastore.Read / Datastore.Write	Damage data	X	X	X	X	X
	ARPCachePoisoning	ARP Cache Poisoning	X	X	X		
Host	UserAccess / AccessControl.Bypass	O	Privilege Escalation	X	X	Host Compromise	
	PrivilegeEscalation						
	Compromise						
	PhysicalZone.Compromise / USBAccess / PhysicalAccess	PhysicalZone Compromise	X	X	X		
	UserAccess	Access to Host		Exploit vulnerabilities and compromise Host	X		
	SoftwareProduct.FindExploit / DevelopExploit / DevelopZeroDay / FindEndOfLifeVuln						
	FindExploit	O	Find exploit				
	DeployExploit	O	Deploy exploit				
	BypassAntiMalware	O					
	Compromise	O	Compromise Host	X	X		
Service	Connect / AccessControl.NonRootLogin	Non-root shell login to Service	X	X	Service Compromise		
	NonRootShellLogin						
	Host.UserAccess						
	UserAccess	Access to Service		Exploit vulnerabilities and compromise Service			
	SoftwareProduct.FindExploit / DevelopExploit / DevelopZeroDay / FindEndOfLifeVuln						
	FindExploit	O	Find exploit				
	DeployExploit	O	Deploy exploit				
	BypassAntiMalware	O					
	Compromise	O	Compromise Service	X			
	Connect / AccessControl.RootLogin	Root shell login to Service / Host	X	X			
	RootShellLogin						
	Service(non-root).Host.UserAccess / Host						
Router	Firewall.DiscoverEntrance / Forwarding	Router Forwarding	X	X	X	X	X
Datastore	Write / Delete	Damage data	X	X	X	X	X

* Steps in grey are optional in the aggregation

Asset	Attack Step	Attack Steps Based Abstraction			Assets Based Abstraction		
		1st Level	2nd Level	3rd Level	4th Level	5th Level	6th Level
SoftwareProduct	FindPublicPatchableVulnerability / FindPublicUnpatchableVulnerability; DevelopExploitForPublic PatchableVulnerability	Deploy Exploit for Public Patchable Vulnerability	×	×	×	×	×
	FindPublicUnpatchableVulnerability / FindPublicPatchableVulnerability; FindExploitPublicUnpatchableVulnerability /FindExploitPublicPatchableVulnerability; *Host./Service./Client.UserAccess*	Find Exploit for patchable / unpatchable vulnerbility	×	×	×	×	×
	DevelopZeroDay	Develop Zero Day	×	×	×	×	×
	FindPublicUnpatchableVulnerability; DevelopExploitForPublicUnpatchableVu lnerability	Deploy Exploit for Unpatchable Vulnerbility	×	×	×	×	×
Web Application	*DiscoverNewVulnerability*; *Service connect*; BypassWAFViaCI; ExploitCI	Exploit Command Injection	Exploit web application	×	×	×	×
	DiscoverNewVulnerability; *Service connect*; BypassWAFViaRFI; ExploitRFI	Exploit Remote File Inclusion					
	DiscoverNewVulnerability; *Service connect*; BypassWAFViaSQLInjection; ExploitSQLi	Exploit SQL Injection					
	DiscoverNewVulnerability; *Service connect*; BypassWAFViaXSS; ExploitXSS	Exploit Cross Site Scripting					

* Steps in grey are optional in the aggregation

References

1. Artz, M.L.: NetSPA: A Network Security Planning Architecture. Massachusetts Institute of Technology, Department of Electrical Engineering and Computer Science (2019)
2. Ekstedt, M., Johnson, P., Lagerström, R., Gorton, D., Nydrén, J., Shahzad, K.: Securicad by foreseeti: a cad tool for enterprise cyber security management. In: 2015 IEEE 19th International Enterprise Distributed Object Computing Workshop, pp. 152–155. IEEE (2015)

3. Homer, J., Varikuti, A., Ou, X., McQueen, M.A.: Improving attack graph visualization through data reduction and attack grouping. In: Goodall, J.R., Conti, G., Ma, K.-L. (eds.) VizSec 2008. LNCS, vol. 5210, pp. 68–79. Springer, Heidelberg (2008). https://doi.org/10.1007/978-3-540-85933-8_7

4. Hong, J., Kim, D.: HARMs: hierarchical attack representation models for network security analysis. In: Australian Information Security Management Conference, p. 12 (2012)

5. Hong, J.B., Kim, D.S., Chung, C.J., Huang, D.: A survey on the usability and practical applications of graphical security models. Comput. Sci. Rev. **26**, 1–16 (2017)

6. Johnson, P., Lagerström, R., Ekstedt, M.: A meta language for threat modeling and attack simulations. In: Proceedings of the 13th International Conference on Availability, Reliability and Security, p. 38. ACM (2018)

7. Johnson, P., Vernotte, A., Ekstedt, M., Lagerström, R.: pwnPr3d: an attack-graph-driven probabilistic threat-modeling approach. In: 2016 11th International Conference on Availability, Reliability and Security (ARES), pp. 278–283 (2016)

8. Kaynar, K.: A taxonomy for attack graph generation and usage in network security. J. Inf. Secur. Appl. **29**, 27–56 (2016)

9. Kotenko, I., Stepashkin, M.: Attack graph based evaluation of network security. In: Leitold, H., Markatos, E.P. (eds.) CMS 2006. LNCS, vol. 4237, pp. 216–227. Springer, Heidelberg (2006). https://doi.org/10.1007/11909033_20

10. Li, E., Barendse, J., Brodbeck, F., Tanner, A.: From A to Z: developing a visual vocabulary for information security threat visualisation. In: Kordy, B., Ekstedt, M., Kim, D.S. (eds.) GraMSec 2016. LNCS, vol. 9987, pp. 102–118. Springer, Cham (2016). https://doi.org/10.1007/978-3-319-46263-9_7

11. Mao, X.: Visualization and natural language representation of simulated cyber attacks. Master's thesis, KTH Royal Institute of Technology (2018)

12. MITRE. About ATT&CK (2018). https://attack.mitre.org/. Accessed 01 Apr 2019

13. MITRE. About CAPEC (2018). https://capec.mitre.org/about/index.html. Accessed 25 Mar 2019

14. Noel, S., Harley, E., Tam, K.H., Limiero, M., Share, M.: Chapter 4 - cygraph: graph-based analytics and visualization for cybersecurity. In: Cognitive Computing: Theory and Applications, volume 35 of Handbook of Statistics, pp. 117–167. Elsevier (2016)

15. Noel, S., Jajodia, S.: Managing attack graph complexity through visual hierarchical aggregation. In: Proceedings of the 2004 ACM Workshop on Visualization and Data Mining for Computer Security, pp. 109–118. ACM (2004)

16. Sommestad, T., Ekstedt, M., Holm, H.: The cyber security modeling language: a tool for assessing the vulnerability of enterprise system architectures. IEEE Syst. J. **7**(3), 363–373 (2013)

High-Level Automatic Event Detection and User Classification in a Social Network Context

Fabio Persia[1]([✉])(iD) and Sven Helmer[2](iD)

[1] Free University of Bozen-Bolzano, Piazza Domenicani 3, 39100 Bolzano, Italy
fabio.persia@unibz.it
[2] University of Zurich, Binzmühlestrasse 14, 8050 Zurich, Switzerland
helmer@ifi.uzh.ch

Abstract. We present a framework for high-level automatic event detection and user classification in a social network context based on a novel temporal extension of relational algebra, which improves and extends our earlier work in the video surveillance context. By means of intuitive and interactive graphical user interfaces, a user is able to gain insights into the inner workings of the system as well as create new event models and user categories on the fly and track their processing through the system in both offline and online modes. Compared to an earlier version, we extended our relational algebra framework with operators suited for processing data from a social network context. As a proof-of-concept we have predefined events and user categories, such as spamming and fake users, on both a synthetic and a real data set containing data related to the interactions of users with Facebook over a 2-year period.

Keywords: Event query languages · High-level event detection · Intervals · Social network analysis · Behavior identification in OSNs

1 Introduction

In the era of big data we have to cope with continuously increasing data collections and data streams originating from various sources. Here, we look at logs containing user interactions with a social network. Usually, the persons (or systems) evaluating the data are not interested in looking at the enormous amount of raw events, but want to be informed about events on a more abstract level. For instance, a log of user activities may pick up every single click of a user, but when investigating the data we may be much more interested in detecting potentially harmful activities and user categories, such as *spamming* or *fake users*.

In fact, Benevenuto et al. [3] analyze user activities through *clickstream data*, initially focusing on statistical properties related to traffic and session workloads

This work was supported by an internal grant from the Free University of Bozen-Bolzano under IN2078 (HAMSIK - High-level AutoMatic event detection in a SocIal networK context).

© Springer Nature Switzerland AG 2019
M. Albanese et al. (Eds.): GraMSec 2019, LNCS 11720, pp. 203–219, 2019.
https://doi.org/10.1007/978-3-030-36537-0_10

(e.g. access frequency, session duration, etc.). In a second phase, they employ a first-order Markov Chain to define a model of behavior describing dominant activities and transition rates between them. Schneider et al. [12] perform an analysis of clickstream data to identify typical user navigation strategies. They reconstruct clickstreams from anonymous HTTP header traces obtained from passively monitored network traffic with thousands of users from different Internet Service Providers and then apply a flexible methodology for identifying user sessions within the OSN. In addition, Amato et al. present a two-stage method for anomaly detection in the behavior of persons while using a social network [1,2]. In a first step, a Markov chain model is used to automatically learn typically normal behavior of users. In a second step, an activity detection framework based on a possible worlds model is applied to detect unexplained activities deviating from the normal behavior.

Our approach models user behavior using complex events. Complex events are usually described in terms of individual simple events standing in a certain temporal relationship with each other. At the core of our system is a temporal relational algebra used to process high-level (and also medium-level) events. This allows us to tap into new efficient methods developed for processing data in temporal database systems [4,11]. We deal with the complexity of high-level events by dividing our system into three layers. The lowest layer generates raw events, in our case related to individual time-stamped observation data depicting users' interactions with a social network (e.g., Facebook or Twitter). This layer is highly dependent on the application domain and has to be adapted if we want to move to another domain (we started from a video surveillance context [8], but we are also planning to apply our framework to data from other heterogeneous data sets, such as *Wikipedia, Yago*[1], or *GovTrack*[2]). The middle layer takes raw events and creates simple events whose format is largely independent of the application domain, thus separating the high-level event detection from technical details of the raw events. Additionally, events generated by the middle layer already contain some aggregated data, simplifying the high-level detection. Finally, on the highest layer a user can construct the complex events that they are really interested in, using medium-level events as building blocks. For ease of use, we also provide a graphical user interface (GUI) for formulating high-level events. The motivation of this paper comes from the fact that, to the best of our knowledge, there are no other interactive frameworks in the social network analysis context able to carry out the overall monitoring process from the lowest up to the highest layers, as well as to improve the support users receive in defining the event models they want to look for by means of smart graphical user interfaces.

As a result, a user can employ our system in a highly interactive way. In our demo, all the different parts of the event detection process on all three layers of the system can be observed in action and also be modified. Event detection

---

[1] http://www.mpi-inf.mpg.de/departments/databases-and-information-systems/research/yago-naga/yago/.

[2] https://www.govtrack.us/.

can be run in two different modes. In the *offline* mode, a historical data set is analyzed after all the raw events have been generated and are stored, for example in a database. This is generally used for forensic purposes. In the *online* mode, the time-stamped low-level data is immediately processed by the system as it is generated. For our demo, we plan to use two data sets - a synthetic and a real one - that can be analyzed in the offline and online mode.

In summary, we make the following contributions:

- We present all three layers of a highly interactive event detection system, ranging from the generation of raw events to the formulation of complex high-level events.
- The system is based on an extension of relational algebra, ISEQL (Interval-based Surveillance Event Query Language) [6], enriched with powerful temporal operators.
- With respect to its previous version [6], we further enhance the expressivity of ISEQL, by introducing two new operators - *cardinality* and *overlap percentage* - and implementing them in the form of *PostgreSQL* stored procedures; such operators are particularly useful for defining high-level event models and user categories in a social network context.
- In the demo, we show the user interfaces of the system and also provide insights into the inner workings by allowing users to run event detection in an offline as well as an online mode.

2   System Architecture

The overall architecture of our proposed system is shown in Fig. 1. It consists of three layers: an *online social network (OSN) crawler*, an *interval action detector* and a *high-level event detector*. A similar architecture was the topic of earlier work [5,6,8,9] in a video surveillance context. The output of the OSN crawler consists of a set Λ of collected data related to user sessions on a particular OSN. More specifically, examples of such interactions are a user "*FABIO*" who logged in at timestamp "2017-05-17 11:39:12", or a user "*SVEN*", who received a message at timestamp "2017-05-17 11:39:27" (Table 1). The interval action detector extracts medium-level events from Λ by labeling a sequence of OSN log entries with descriptors such as "Status&Friends" or "Shares". This layer produces as output a set M of medium-level annotations referring to intervals of entries within the log. Consequently, each user session at this semantic layer is modeled by means of a sequence of higher level intervals, rather than with a list of time-stamped low-level action symbols. Eventually, the high-level event detector takes a set E of event models and determines whether any of these events occur in M. Moreover, it also performs the user classification, thus assigning to each tracked user a category referring to a specific temporal interval. Thus, we assume the availability of a *log* describing a sequence of user interactions with an OSN. Our aim is to discover subsequences – in a log recording user activities – matching models of known user behavior and to perform an effective user classification (the formal definitions of *OSN Log*, *Interval Labeling*, *High-Level Events*, and *User Classifications* are given in [10]). More specifically, each of the listed layers consists of three different sublayers (Fig. 2): a *graphical user interface (GUI)*, an *application core*, and a *database*. The *GUI*, implemented using *Java Swing APIs*, allows users to interact intuitively with the framework, guiding them along in a step-by-step manner, making our system usable for people with no prior knowledge in relational algebra. The *Application Core*, developed in Java, collects all the input coming from the *GUI*, checking it for correctness. On the other end, it stores *PL/pgSQL* versions of *medium-level* and *high-level* event models to make them persistent. It also invokes existing event models on a specified data set, detecting the specified events in the data set. *PostgreSQL 9.4* is the underlying *database* and every operator - including *cardinality* and *overlap percentage* - of both a *medium-level* and a *high-level* event model is implemented via stored procedures in a PostgreSQL DBMS. The individual operators can be assembled dynamically into different event models.

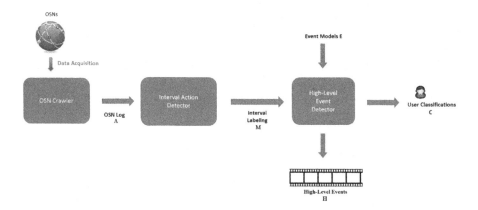

Fig. 1. Overall architecture

Fig. 2. Semantic sub-layers

2.1 OSN CRAWLER

In an earlier project, an OSN crawler to collect data from Facebook users was developed [1]. More specifically, this was done with PHP utilizing Facebook APIs to capture data within secure sessions following the OAuth protocol. The application was then shared on Facebook and collected, after receiving authorization from users, data related to the interactions of about 1600 users over a two-year period from 2013 to 2015. The collected data was anonymized by assigning a random ID to each user and dropping personal details such as age and gender. Additionally, all participating users were informed that our work was purely research-related and that there were no commercial uses (falling under Facebook's privacy policy, which prohibits and punishes unlawful misuse). In the long run, the aim of the research is to protect users from fraudulent behavior. We are aware that technology like this can be misused, but this is true for security-related methods in general and needs to be discussed in a wider context [7].

Table 1 shows an example of a log obtained by the OSN crawler from Facebook (for the sake of readability we have replaced the random user IDs with our own names in this and the following tables). Table 2 gives an overview of the atomic user actions (along with their high-level categories) that we captured. Clearly, the OSN Crawler can also be used for capturing user interactions with other OSNs, such as *Twitter*; in that case, the content of the user tweets can also be collected and then stored.

Table 1. Example of OSN Log

Action symbol	User	Timestamp	IP
login	FABIO	2017-05-17 11:39:12	192.168.1.88
likes a page	FABIO	2017-05-17 11:39:20	192.168.1.88
login	SVEN	2017-05-17 11:39:24	
message sent	FABIO	2017-05-17 11:39:27	192.168.1.88
message received	SVEN	2017-05-17 11:39:27	
status wall post	SVEN	2017-05-17 11:39:30	
message sent	SVEN	2017-05-17 11:39:40	
message received	FABIO	2017-05-17 11:39:40	192.168.1.88
logout	SVEN	2017-05-17 11:39:42	
logout	FABIO	2017-05-17 11:39:50	192.168.1.88

2.2 Interval Action Detector

The task of the *Interval Action Detector* is to assemble individual time-stamped action symbol with low-level labels into meaningful events described by an interval. In this way, each user session is modeled by means of a sequence of higher

Table 2. Facebook predicates and related action symbols in Facebook logs

Category	Atomic action
login	login
status&friends	status wall post
	friend approved
	mobile status update
	checkin
	status update
messages	message received
	message sent
photos	added picture
	tagged in a picture
shares	youtube video shared
	youtube created story
	link app created story
	published link
	link shared story
	video shared story
	pictured shared story
like	likes a page
logout	logout

level intervals, rather than with a list of time-stamped low-level action symbols. More specifically, the *Interval Action Detector* aggregates the results from the OSN Crawler into medium-level events using medium-level predicates corresponding to the categories defined in [3] and shown in Table 2. Additionally, Table 3 shows the Interval Labeling obtained by processing the OSN Log shown in Table 1. In order to do that, we utilize interval-based extensions we have introduced in our earlier work on the detection of high-level events [5,6]. All operators, including the new ones for analyzing social network data, were implemented in PostgreSQL as stored procedures. We do not describe our extensions in detail, but introduce the concepts as needed (for details of our interval-based language, see [5,6]).

The *Interval Action Detector* works both in an *offline* mode - where it has access to a complete (historical) data set - and in an *online* mode - where it works similar to a continuous query in a data stream management system, thus being able to react in real time. In addition, the GUI also allows to broaden both the category and the atomic action sets (Table 2), as well as to select just a subsets of the categories to be detected by the Interval Action Detector (by default, all of them are searched).

Table 3. Example of interval labeling

Pred	Start	End	Arg_1
session	17-05-17 11:39:12	17-05-17 11:39:50	FABIO
login	17-05-17 11:39:12	17-05-17 11:39:12	FABIO
like	17-05-17 11:39:13	17-05-17 11:39:20	FABIO
messages	17-05-17 11:39:21	17-05-17 11:39:40	FABIO
session	17-05-17 11:39:24	17-05-17 11:39:42	SVEN
login	17-05-17 11:39:24	17-05-17 11:39:24	SVEN
messages	17-05-17 11:39:25	17-05-17 11:39:27	SVEN
status&friends	17-05-17 11:39:28	17-05-17 11:39:30	SVEN
messages	17-05-17 11:39:31	17-05-17 11:39:40	SVEN
logout	17-05-17 11:39:42	17-05-17 11:39:42	SVEN
logout	17-05-17 11:39:50	17-05-17 11:39:50	FABIO

2.3 High-Level Event Detector

The main goal of the high-level event detector is to combine medium-level events into descriptions of complex events. This is done by putting the intervals associated with two or more medium-level events in relation to each other, thus exploiting the semantic interval relationships defined in [5] and used in *ISEQL* [6]. In principle, we have five different operators, here visualized by a small sketch indicating the relative position of two intervals: LEFT OVERLAP (⇌), DURING (⇌), START PRECEDING (⇌), END FOLLOWING (⇒), and BEFORE (⁃⁃). All of these relations also have a reverse counterpart: RIGHT OVERLAP (⇌), REVERSE DURING (⇌), REVERSE START PRECEDING (⇌), REVERSE END FOLLOWING (⇌), and AFTER (⁃⁃).

The constructs employed by ISEQL are not just simple Allen relationships, but we have extended and parameterized them to match the specific needs of event detection. For example, the relation BEFORE covers "takes place before" and "meets". The parameter δ of the BEFORE relation can be used to tune the desired maximum distance between the intervals, meaning that $\delta = 0$ models "meets", all other values model "takes place before".

We introduced two new constraints – the *cardinality* and *overlap percentage* [10], which allow us to formulate high-level event models in social network environments more easily, thus also enhancing the expressivity of ISEQL. These new operators were also integrated into the GUI.

For instance, we can utilize these new operators and constraints to define a *spamming* event model. We interpret spamming as sharing something more than k times *(Case 1,* Fig. 3*)* or spending more than p percent of a session with sharing activities *(Case 2,* Fig. 4*)*. In order to model and then detect such scenarios, we need to combine different intervals, each of them corresponding to a medium-level event. More specifically, for modeling *Case 1* we exploit the *right cardinality (k)* constraint [10] between "session" and "shares" intervals, which guarantees that there are at least k "shares" intervals within the same "session". On the other hand, for modeling *Case 2* we use the *left overlap percentage* constraint [10] between "session" and "shares" intervals, which states that at least p percent of the session was spent by the user *sharing* something. Figures 3 and 4 show examples of instances that would be classified as spamming: Fig. 3 for $k = 3$ (Case 1), and Fig. 4 for $p = 0.8$ (Case 2).

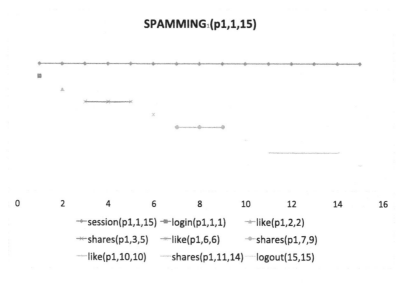

SPAMMING₁(p1,1,15)

Fig. 3. Spamming - Case 1, $k = 3$

While an experienced user can define high-level events directly in ISEQL, for an ordinary user this task may be too daunting. With the help of our GUI, a user is guided through the steps of defining a high-level event. This includes drawing intervals on a canvas and supplying parameters. In the background, the system checks the model for consistency, transforms it into a temporal relational algebra expression, and generates the code in form of *PL/pgSQL* for the actual event detection. As an example, Figs. 5 and 6 depict how the *spamming* high-level event (see Figs. 3 and 4) is entered using the GUI. More specifically, it is defined as logical disjunction of *Case 1* (Fig. 3) and *Case 2* (Fig. 4). Consequently, by answering some simple questions (Fig. 5)[3], and drawing the desired relationships among intervals on a temporal canvas (Fig. 6), the user is able to effectively define a high-level event model. The black intervals in Fig. 6 represent combinations of simpler events and are automatically generated by the system. This *high-level* event can now be stored, queried, and re-used as a building block to define more complex events. More details are provided in Appendix A.

By default, our queries work at a *global granularity*, looking at all sessions of all users. However, we can also run queries at a finer granularity, defining

---

[3] We also provide an online help.

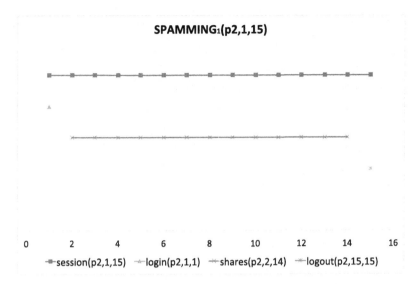

Fig. 4. Spamming - Case 2, $p = 0.8$

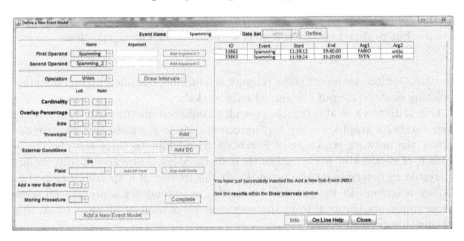

Fig. 5. High-Level Event Detector

specific temporal intervals for users. We can even analyze a user's behavior by investigating and categorizing each of their sessions individually. The session type that appears most frequently is then used to classify a user. For example, a user most of whose sessions are marked as *inactive* - that means that they are not an instance of any of the searched event models - is classified as a *fake*

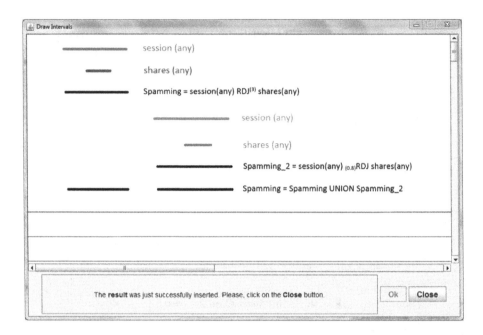

Fig. 6. High-Level Event Detector - intervals on a temporal canvas

user. In addition, we report some relevant screenshots classified by functionality, depicting how the overall system actually works[4].

Our framework is also flexible enough to enhance queries with data coming from Facebook graphs, adding further constraints to them. In this way, we can exploit the network structure of Facebook friendships to carry out the classification of *clusters of users* rather than single users. More specifically, we are interested in determining the category all the friends of a specific user u (i.e., his/her neighbors in the graphs) belong to, or extend the search to all users whose minimum distance from the user u is lower than or at most equal to a specific threshold d.

2.4   Graphical User Interface (GUI)

As shown in Fig. 2, each of the semantic layers of Fig. 1 consists of a *graphical user interface (GUI)*, an *application core*, and a *database*. Since many of the framework users may be inexperienced, both in social network analysis and in relational algebra, it may be non-trivial for them to define tasks and models. Thus, the GUI plays an important role in helping them to interact effectively with the framework and to make use of specific functionalities, such as the *definition of a new high-level event model.* More details about the functionalities of the

---

[4] https://www.dropbox.com/sh/um0yucb8810nrhu/AAAt5kbr9Tsz4moEgghKgxeja?
dl=0.

framework and the way to utilize them via the graphical user interfaces are exhibited in Appendix A.

3   Demo Specifications

For the purpose of our demo, we use two different data sets: the Facebook data set we already mentioned earlier [1] and a synthetic data set produced by a generator. The size of the data set and the density of the events (i.e., the number of events per time unit) can be controlled via parameters by a user. We decided to use Facebook in our demo, since it is still the world's most popular social network; however, the *OSN Crawler* could potentially work also on other OSNs, such as Twitter, by using the Twitter APIs with the default access level.

We plan to start the demo by detecting high-level events on one of the above social data sets in offline mode to illustrate how the event detection works in principle. This involves identifying events such as different scenarios of potential spamming in the social log. However, depending on the particular interests of a demo participant we can either focus our attention on one specific layer (*OSN Crawler, Interval Action Detector*, and High-Level Event Detector) or carry out the whole process from the low-level label extraction all the way to the detection of high-level events. For the online mode, we stream one of the data sets past the event detector, emitting the atomic events according to their timestamps.

Participants will be able to discover some of the predefined medium-level events, investigate the interval labeling, or create new medium-level events and detect them afterwards. On the top-most layer, users can do the same for high-level events: detecting them and defining new ones.

All the functionality of every layer is accessible via intuitive GUIs that provide continuous feedback about what is happening in the system. For the interval action and high-level event detectors, a participant is not only able to define new events and detect them, but they can also have a closer look at the interval representation of events in the form of stored procedures, and define and detect them on the fly. They can also investigate the global behavior of a user over a specified temporal interval, thus obtaining the category of the user.

4   Conclusion and Future Work

In this paper, we present a smart and interactive framework for automatic event detection and user classification in a social network context. More specifically, the user is able both to easily define high-level event models by a means of a smart graphical user interface, and to discover their instances in a real data set containing data dealing with interactions of users with Facebook, as well as in synthetic data sets.

Future work will be devoted particularly to further enhance the framework efficiency. In fact, we plan to develop a family of efficient plane-sweeping interval join algorithms that can evaluate the wide range of interval relationships predicates defined in [6], that are broadly exploited by our framework, directly in the query processing framework of the *PostgreSQL* DBMS. These predicates also include the *cardinality* and *overlap percentage* operators, that are particularly useful for modeling high-level events and user categories in a social network context. Additionally, we also plan to improve the framework response time for use in live data streams. This involves the development and the dissemination of a specifically designed and developed web application able to capture user interactions with a social network in real time, compatibly with its privacy policy.

Appendix

A   The Framework Functionalities

In this appendix we take a closer look at the main functionalities provided by the framework for high-level automatic event detection and user classification. More specifically, we list them below and give more details in the following sections.

- Detection of Low-Level Annotations (Sect. A.1).
- Detection of Medium-Level Annotations (Sect. A.2).
- Detection of High-Level Event Occurrences (Sect. A.3).
- Detection of User Classifications (Sect. A.4).
- Automatic High-Level Event Detection (Sect. A.5).
- Definition of a New Atomic Predicate (Sect. A.6).
- Definition of a New Medium-Level Predicate (Sect. A.7).
- Definition of a New High-Level Event Model (Sect. A.8).

A.1   Detection of Low-Level Annotations

This functionality allows to import all the low-level annotations occurring within a specified temporal window. So far, they can be imported from two different sources. These annotations are used as input for further processing steps. The first one is a real data set containing data related to the interactions of users with Facebook over a 2-year period, previously collected for [2]. The second one is a synthetic data set, generated by a specifically designated tool, whose size and the density of the events (i.e., the number of events per time unit) can be controlled via parameters by users. However, the system is flexible enough to easily allow in the future imports from other sources, including live data streams, compatibly with the related privacy policy.

A.2   Detection of Medium-Level Annotations

This functionality detects the interval labeling corresponding to the captured *OSN Log* (Fig. 1). More specifically, Fig. 7 shows the medium-level annotations corresponding to the OSN Log listed in Table 1. In Fig. 7 we use as *source* the synthetically generated data set *unibz* mentioned in Sect. A.1 and the medium-level predicates *status&friends*, *messages*, *photos*, *session*, *shares*, *like*, *logout* in *offline* mode. The collected interval labeling is shown on the right-hand side of Fig. 7 and can be further processed in order to infer both high-level events and user classifications.

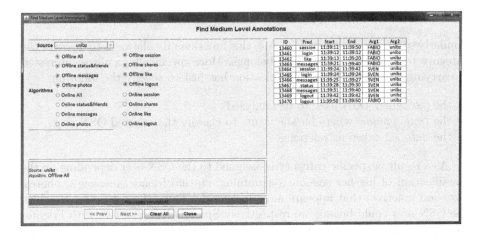

Fig. 7. Detection of Medium-Level Annotations

A.3 Detection of High-Level Event Occurrences

This functionality detects occurrences of high-level events whose models are stored in the knowledge base. In this framework the knowledge base of event models is stored as set of stored procedures in the *PostgreSQL* database management system. As shown in Fig. 8, the user simply needs to select the *event* to be discovered (*SPAM* in this case), and the data set to be investigated (*unibz* in this case). Clearly, for a data set to be available, it has to be first processed, i.e., it has to be labeled via *interval labeling*). The result of the use case shown in Fig. 8 is an instance of the *SPAM* event detected for the user *FABIO* from 11:39:12 to 19:40:00.

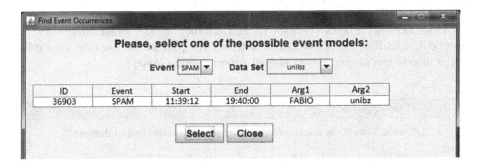

Fig. 8. Detection of High-Level Event Occurrences

A.4   Detection of User Classifications

Similarly to the procedure in Sect. A.3, this functionality allows us to discover the category to which each OSN user belongs. More specifically, a client interested in carrying out an OSN user classification has just to specify the following:

- the particular *OSN user* to be analyzed;
- the *time window* where he/she wants to classify the selected OSN user;
- the *data set* taken as reference.

As a result, a specific category is assigned to the OSN user depending on the classification of his/her sessions (spamming, status&friends, messages, photos, like, and inactive) that appears most frequently. Thus, the categories to which the OSN user could belong are respectively Spammer, Interactive with Friends, Message Sender, Photo Poster, Like Adder, and Fake User. This is due to the fact that all the defined event models are flexible, so they can be also applied to classify users themselves, thus working at a *lower (user) granularity*.

A.5   Automatic High-Level Event Detection

This functionality allows to automatically carry out the overall process described in Sects. A.1, A.2, A.3, and A.4. As a result, the user just needs to specify all the inputs necessary in the previous sections once, and the whole process shown in Fig. 1 is performed; consequently, the output are the high-level events and the user classifications satisfying the inserted constraints.

The process can be run in both offline and online modes. For the online mode, we stream one of the data sets past the event detector, emitting the atomic events according to their timestamps.

A.6   Definition of a New Atomic Predicate

This functionality allows the user to add another atomic event to the set of atomic actions listed in Table 2. For instance, the user in the use case shown in Fig. 9 inserts the atomic action named *Interact with Game*.

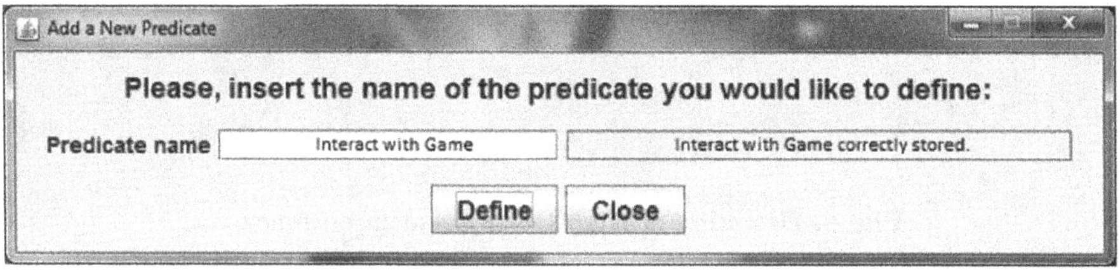

Fig. 9. Detection of High-Level Event Occurrences

A.7   Definition of a New Medium-Level Predicate

Similarly to adding atomic predicates as described in Sect. A.6, this functionality allows us to insert a new medium-level predicate into the set of categories listed in Table 2. More specifically, by means of another smart graphical user interface, the user is able to directly write the PL/pgSQL code of the new medium-level predicate, also specifying the relationships with the low-level atomic actions.

A.8   Definition of a New High-Level Event Model

As mentioned in Sect. 2, this functionality allows a user who is not familiar with relational algebra to easily define a high-level event model; Figs. 5 and 6 illustrate an example for using the smart graphical user interface for defining the *Spamming* event model.

In order to illustrate the advantages of the user interface, we describe the procedure for defining a *new event model* in the following. This is done in a step-by-step manner, by asking the user for (see Fig. 5):

- the *name* of the new event (field *Event Name*);
- the *data set* he or she would like to explore (from a list of available data sets) (field *Data Set*);
- the *medium-level predicates* (or, as an alternative, already-defined events) associated with the intervals (operands) that he or she is currently adding to the *global event* (fields *First Operand, Second Operand*);
- optional values for the *arguments* of the *first/second operand* in case of a medium-level predicate (field *Argument*, close to *First/Second Operand*); arguments can be easily added by clicking on the *Add Argument* button;
- the possibility to carry out set operations between the two inserted interval predicates (field *Operation*);
- drawing the two intervals (after clicking on the *Draw Intervals* button); then, the *application core* will capture the values of the *left* and *right endpoints of both intervals* (see for instance first and second lines of Fig. 6);
- specifying how often the left/right interval (fields *Left/Right Cardinality*, respectively) has to appear in the result set. If the user selects *YES*, a pop-up window will ask to select among three options; at least k times (k to be specified), more than one tuple (*), or exactly one tuple (*one*) [10]; otherwise, no further constraints are added;
- specifying the overlap percentage between the two intervals with respect to the left/right interval (fields *Left/Right Overlap Percentage*, respectively). In case the user selects *YES*, a pop-up window will ask for the overlap percentage (from 0% to 100%); otherwise, no further constraints are added;
- whether he or she wants to take into account the *relationships* between the left/right endpoints (fields *Left Side, Right Side*);
- the maximum distance between interval endpoints (fields *Left/Right Threshold*); in case of *overlapping events* checking whether to take into account the distance between the *left endpoints* of the *first* and *second operand* or between

the *right endpoints* of the two *operands*. In case of *non-overlapping events*, a user has to specify whether to take into account the distance between the *right endpoint* of the *first operand* and the *left endpoint* of the *second operand* or between the *left endpoint* of the *first operand* and the *right endpoint* of the *second operand*. Depending on the information provided by the user, the application core infers the specific operator that will be applied.

– the optional *additional constraints* between the *first* and *second interval* he or she would like to add, starting from the *partial result set* (clicking on *Add EC*, close to the *External Conditions* field, and then allowing the addition of constraints via a mask);
– the *fields* he or she would like to project with reference to the current result set (field *Field*); the user just needs to select the fields to be projected, and click on *Add i-th Field*;
– whether he or she wants to add *more intervals* to the *complex event* he or she is defining (field *Add a new Sub-Event*); in that case, the process is repeated starting from the third bullet point;
– whether he or she wants to store the *event model* as a *PL/pgSQL procedure* (field *Storing Procedure*).

After each step the *application core* checks the consistency of the input. At the end of the procedure, a summary with the retrieved instances, if any, will be visible to the user.

References

1. Amato, F., et al.: Recognizing human behaviours in online social networks. Comput. Secur. **74**, 355–370 (2018)
2. Amato, F., De Santo, A., Moscato, V., Persia, F., Picariello, A.: Detecting unexplained human behaviors in social networks. In: Proceedings of the 2014 IEEE International Conference on Semantic Computing, ICSC 2014, pp. 143–150. IEEE, Newport Beach (2014)
3. Benevenuto, F., Rodrigues, T., Cha, M., Almeida, V.: Characterizing user navigation and interactions in online social networks. Inf. Sci. **195**, 1–24 (2012)
4. Dignös, A., Böhlen, M., Gamper, J.: Overlap interval partition join. In: International Conference on Management of Data, SIGMOD 2014, pp. 1459–1470. ACM, Snowbird (2014)
5. Helmer, S., Persia, F.: High-level surveillance event detection using an interval-based query language. In: Proceedings of 2016 IEEE International Conference on Semantic Computing, ICSC 2016, pp. 39–46. IEEE, Laguna Hills (2016)
6. Helmer, S., Persia, F.: ISEQL: an interval-based surveillance event query language. Int. J. Multimed. Data Eng. Manag. (IJMDEM) **7**(4), 1–21 (2016)
7. Irwin, A.S.M.: Double-edged sword: dual-purpose cyber security methods. In: Prunckun, H. (ed.) Cyber Weaponry. ASTSA, pp. 101–112. Springer, Cham (2018). https://doi.org/10.1007/978-3-319-74107-9_8
8. Persia, F., Bettini, F., Helmer, S.: An interactive framework for video surveillance event detection and modeling. In: Proceedings of the 2017 ACM on Conference on Information and Knowledge Management, CIKM 2017, pp. 2515–2518. ACM, Singapore (2017)

9. Persia, F., Bettini, F., Helmer, S.: Labeling the frames of a video stream with interval events. In: Proceedings of the 2017 IEEE International Conference on Semantic Computing, ICSC 2017, pp. 204–211. IEEE, San Diego (2017)
10. Persia, F., Helmer, S.: A framework for high-level event detection in a social network context via an extension of ISEQL. In: Proceedings of the 2018 IEEE International Conference on Semantic Computing, ICSC 2018, pp. 140–147. IEEE, Laguna Hills (2018)
11. Piatov, D., Helmer, S., Dignös, A.: An interval join optimized for modern hardware. In: Proceedings of the 2016 IEEE 32nd International Conference on Data Engineering (ICDE), pp. 1098–1109. IEEE, Helsinki (2016)
12. Schneider, F., Feldmann, A., Krishnamurthy, B., Willinger, W.: Understanding online social network usage from a network perspective. In: Proceedings of the 9th ACM SIGCOMM Conference on Internet Measurement, pp. 35–48. ACM, New York (2009)

Author Index

Printed in the United States
By Bookmasters